藍海文化

Blueocean

www.blueocean.com.tw

教學啟航 · 知識藍海

目錄

序

這是一個發展中的課題，且越來越重要：

聯合國的永續發展目標 SDGs 17 尚未收官驗收，因應 COVID-19 的聯合國的 2030 永續發展議程如同新世界的社會契約，對應 18 世紀盧梭的社會契約，自由人的定義衍生到數位世界與各方需求，發展出 SDG18 的想像框架，如數位科技合作、外太空、寮國目標等。

SDG18 數位科技合作篇認為：個人權利、社會平等和民主的威脅，而這些風險和不確定性都被「數位落差」（Digital Divide）更加倍放大了，企業的企業社會責任（CSR），在數位科技為基之下，用永續經營（ESG）精神與全球報導倡議機構（GRI）準則進行企業社會責任的佈建，不久的未來有機會成為另一個財務上堪比每股盈餘（EPS）的重要指標，社會責任投資（SRI）的課題也因此更加成為顯學，而這樣的安排，恰恰滿足了第二部門私企業以營利最大化為目的的企業最高指導方針，同時兼顧了企業在社會上應負的社會責任，是一個對人類永續發展有益與切實可行的好策略。

發展 CSR 的課題，是一個摸著石頭過河的過程，臺灣目前尚少見本土化的實務教材，多數是發展中的課程簡報。如今，欣見《薪傳智庫 CSR 實務商模戰略》一書的出版，它代表了臺灣在 CSR 的領域上已然呈現本土扎根的真實痕跡，並且正嵌入在追求永續發展的企業 DNA 中。如此反身性發展（reflexive development）的企業活動，說明了原先歐洲福利國家對於人民責任的社會、經濟、環境、文化等四個支柱，隨著時間的變遷，進化到自資本主義到博愛主義的經濟系統光譜中，已逐漸看見了人類經濟史上很久以來就缺乏的社會關注。因此從聯合國 SDGs、ESG、公共財、外部效果的公共事務相關學理與實務中、在在都可以發現如同諾貝爾經濟學獎得主 E. Ostrom 提到的公共池的治理，同時也看見發展企業社會責任的可行解與相關工具正被應用在臺灣的本地發展上，如同本書所言：我們得先學習如何先成為 B 型企業，重視社會投報率（SROI）而非僅是投報率（ROI），發展 CSR 的同時，強化 SRI 的稽核機制與人民共同監督的機制，並研構以 SDG18 為基的 CSR 人才培育機制，相信如此的企業經營途徑（methodology），讓我們在追求企業成長的同時，也能兼顧地球環境的永續，以及人類社會生存的永續課題。

國立高雄科技大學財務管理系兼任助理教授
台灣產學研合作發展策進會協會理事長
教授國立高雄科技大學財務管理系企業社會責任實務課程
國立中山大學公共事務管理研究所博士

黃煒能 博士

資料來源

- Amy Luers（作者）；CSRone 鄭鈺弘（編譯）（2020 年 11 月 22 日）。【CSRone】SDGs 的缺角：數位科技為人類和地球服務 SDG18。取自 https://csr.cw.com.tw/article/41743
- GRI 執行長 Tim Mohin（黃楷元／譯）（2019 年 3 月 14 日）。【CSRone】ESG 趨勢》GRI 執行長 Tim：企業永續將成為主流顯學。取自 https://csr.cw.com.tw/article/40894

一、OECD – CSR

（一）COVID-19 疫情

　　COVID-19 新冠肺炎疫情全球大爆發疫情肆虐，現在全球逾 50 國驗出變種病毒蠢蠢欲動，世界衛生組織（以下簡稱 WHO）表示，有鑑於病毒的傳染模式，2021 年的疫情可能比第一年嚴峻，傳染力將更強且蔓延擴散中；對各國防疫、經濟掀起海嘯般的巨浪，回首 2020 年初的重重挑戰猶如各地方政府的試煉，當然全球及臺灣供應鏈與經濟的影響是史上最強大的，除了造成全球社會與經濟發展腳步放緩外，病毒威脅著全球體系，包含生命消亡、金融崩盤、政治角力的複合式的全球危機，除了帶走人們寶貴的性命、經濟陷入停滯之外，未來政治、社會、經濟市場的「不確定性」將成為常態。肩負臺灣產業發展重責大任的中小企業體，若紛紛不敵疫情衝擊吹起熄燈號，這股浪潮勢必撼動國本並嚴重影響臺灣的發展，包含這場疫情改變了現代人的價值觀，在疫情後的經濟復原規劃中，禍福相依，中小企業應該如何將這場危機轉化為轉機？本書刊關鍵在快速化危機為轉機的應變能力，逆境布局，以商業模式緩解危機；提升營運的可擴展性和靈活性，提供增加經濟和社會利益的最佳方法，放眼後疫情時代的先機和商機，解決不平等的最佳時刻，即使整個世界經濟動盪，臺灣企業依舊儀態萬千也氣勢十足。

　　20 世紀 90 年代開始，全球化和貿易自由化浪潮襲捲全世界，目前全球前十大經濟體一半以上都是跨國企業，企業漸趨以策略的角度關注企業社會責任（以下簡稱 CSR），影響力已經跟國家力量並駕齊驅。而整個疫情與企業風控還需要政策的引導拉高層次，CSR 議題的驅動力已成為一個國家文化、品牌。大數據調查統計超過八成六企業認為，實踐 CSR 是最能提升企業形象，根據研究顯示，組織內所有業務發展價值中，無形資產的佔比高達 80%，顯示企業聲譽對其 CSR 經營發展的重要性。有鑑於此，有不少臺灣中小企業以全球市場為戰場，在全世界各地尋求多元商機，如今 CSR 已經在國際上形成一股很強的力量，各國意識抬頭關鍵更加影響政府與企業行為的焦點趨勢，企業與國家之間應延展互信加強全面與動態的理解合作至關重要，各經濟發展的數據統計：企業推動 CSR 事務時，最大困境竟是難以去評估效益，企業不知道該做什麼與不該做什麼？特別是業界跟政府及學界之間應如何歸屬責任，至今無明確的解答。然而以《OECD 多國企業指導綱領》共有 10 項指導原則，環球議題的 CSR 角度看來，宏觀的 CSR 真諦在每次落實執行與每個行動當下；亦是薪傳智庫數位科技引領數位趨勢善用科技從優化到轉型運營，解決政府機關與中小企業戰略運行的高瞻遠矚，智慧高效能帶給機關與中小企業更多戰情，全球戰略決策價值展現未來新樣貌。

（二）企業有四個不夠

　　全球投資 CSR 已是顯學，面對這波疫情的衝擊，臺灣約 98% 的中小企業可說是首當其衝，集體共同強化政府與企業的競爭能力是點燃薪傳智庫數位科技創新的動能，薪傳智庫數位科技致力協助，承諾以最專業的服務品質提供商業戰略動能；讓有心執行 CSR 的中小企業快速上路，因為 CSR 並不是大企業的專利，臺灣有更多中小企業需要共同參與擴散（中小企業家數佔全體企業的比重為 97.80%），在 2020 年突發不可控的 COVID-19 疫情下，讓薪傳智庫在推動 CSR 相關議題時感受到中小企業幾乎面臨生存不易外，很多企業經營者認為，如果執行 CSR，就意味著必須投入企業不少資源，提高經營成本，讓很多的中小企業退避三舍拒絕著手 CSR。薪傳智庫數位科技洞悉企業針對 CSR 會遇到四個不夠：1. 經費不夠；2. 認知不夠；3. 時間不夠；4. 人力不夠，輔導過程發現中小企業對政府政策相對也非常不暸解，誤解與不曾重視政府的用心，中央與地方積極的執行政策，臺灣每年各部會的預算編列都是致力於扶持中小企業產、銷、人、發、財、數位轉型與新南向政策補助（補貼），礙於中小企業只專注本業的投入，資源受限的中小企業面對充滿不確定、反覆無常的新現實（New Reality），未能全方位的深入暸解與國家政策並肩戰略。

（三）企業健檢與簽證

　　中小企業現有的商業模式並未完全反映外部成本內部化的問題，誰有解方？薪傳智庫透過經驗與中小企業深度溝通，針對中小企業面對的未來挑戰，以「系統性的轉型」取代「漸進式的改變」，運用系統更有效的指標來衡量績效，並建構中小企業聯盟與官、學、研攜手整合缺口，藉由多方資源與各部會科專多元互動的全方位解決方案：讓國家與企業攜手邁進國際的價值觀、影響力在效能焦點下的利益極大化，借力使力探索出 CSR 的核心。目前有近八至九成外商及上市企業，按 CSR 表現選擇供應商，藉由薪傳智庫數位科技 CSR 商業模式整合智慧引導系統，俾使企業更便捷地凝結商業戰略，解決產業人力、人才、資金、研發、技術、傳承接班、資源媒合、獎勵補助、產業資訊、國際化相關問題，薪傳智庫數位科技客製化提供企業創新營運模式應用實務，落實經營方針規劃 CSR 的具體策略，讓更多中小企業透過獲得申請國家政策、中央計畫、地方補助與數位科技轉型能量。並提供全方位企業健檢 CSR（ESGs）以及數位雲端資訊存摺留存，讓臺灣政府與中小企業共好的體質持續發展獲利；並透過第三方公正機構：聯合律師以及會計師做年度 CSR 報告背書簽證，讓一個支點帶動加乘的國際經濟效益，積極打造創新的價值鏈產業環境，邁向全球化的科技日新月異與市場無遠弗屆，薪傳營運模式將接續制定短、中、長期的戰略目標邁向企業永續發展界定，聚焦廣為接受的共識性定義，創造共享的核心價值。

資料來源
· 花佳正（2021 年 1 月）。2021 全球經濟展望。取自 http://www.cnfi.org.tw/front/bin/ptdetail.phtml?Part=magazine11001-610-2
· 林雅惠（2006）。臺灣企業社會責任發展之現況（碩士論文）。取自 https://www.airitilibrary.com/Publication/alDetailedMesh1?DocID=U0006-0802200612125000
· 張曉雯（譯）；嚴思祺（核稿）（2021 年 1 月 14 日）。新變種病毒更具傳染力 世衛：第 2 年疫情更嚴峻。取自 https://www.cna.com.tw/news/aopl/202101140055.aspx
· BBC News（2020 年 12 月 31 日）。亞洲多地發現新冠變種病毒，印韓日等收緊入境措施。取自 https://www.bbc.com/zhongwen/trad/world-55496215

目標 1　消除各地一切形式的貧窮。

目標 2　消除飢餓，達成糧食安全，改善營養及促進永續農業。

目標 3　確保健康及促進各年齡層的福祉。

目標 4　確保有教無類、公平以及高品質的教育，及提倡終身學習，讓每一個人都有一份好工作。

目標 5　實現性別平等，並賦予婦女權力。

目標 6　確保所有人都能享有水及衛生及其永續管理。

目標 7　確保所有的人都可取得負擔得起、可靠的、永續的，及現代的能源。

目標 8　促進包容且永續的經濟成長，達到全面且有生產力的就業，讓每一個人都有一份好工作。

目標 9　建立具有韌性的基礎建設，促進包容且永續的工業，並加速創新。

目標 10　減少國內及國家間不平等。

目標 11　促使城市與人類居住具包容、安全、韌性及永續性。

目標 12　確保永續消費及生產模式。

目標 13　採取緊急措施以因應氣候變遷及其影響。

目標 14　保育及永續利用海洋與海洋資源，以確保永續發展。

目標 15　保護、維護及促進領地生態系統的永續使用，永續的管理森林，對抗沙漠化，終止及逆轉土地劣化，並遏止生物多樣性的喪失。

目標 16　促進和平且包容的社會，以落實永續發展；提供司法管道給所有人；在所有階層建立有效的、負責的且包容的制度。

目標 17　強化永續發展執行方法及活化永續發展全球夥伴關係。

薪傳響應國際性倡議

目標 18　數位科技為人類和地球謀福祉，正確運用數位科技促進人類和平福祉。

（一）三大面向之社會影響力

1.【趨勢甦醒關鍵能量】

將引領全國中小企業與邁進國際趨勢，從不知道 CSR 到不懂 CSR 跨過不會做 CSR 去認識 CSR，現在企業議合群戰已取代企業單打獨鬥，薪傳積極推動 CSR 觀念並用更創新的行動及策略布局來帶動各產業，積極具體的評估執行方法與產業模型以及學習地圖，讓中小企業與政府機關暨組織共同攜手重視並落實社會責任。

(1) 薪傳智庫系統數位科技平台將引領全國中小企業與國際市場從不知道 CSR 到不懂 CSR 跨過不會做 CSR 去做到 CSR，並彙整提供所有國內外相關的國際 GRI 標準，從 SDGs 到永續經營（以下簡稱 ESG）以及 B Corp 認證。

(2) 建立薪傳知識庫：讓每家企業在知識庫中找到並展現自家 CSR 的獨特性與特定的核心價值，平台可聯網搜尋所有相關的資訊。

(3) FB 教學影音與由淺入深的平台操作，讓智能的學習模式更加具指標性，倡導國家推動政府機構與企業成為最值得社會尊敬的對象。

2.【新經濟模式的關鍵跳躍點】

薪傳透過 25 年以上的經驗，陪伴國內中小企業與國際暨跨國際企業的輔導做深度溝通，用企業語言深入瞭解其真正需求與企業特質善盡強化 CSR 計畫執行，提供中小企業實務解析與資源整合，推動整合產、官、學、研彼此攜手共享共創資源，薪傳提供中小企業「客製化」創新營運模式暨應用實務，讓企業面臨疫情的危機可轉化成更多元的無限商機，企業發展藍圖可營造規劃健全的 CSR 融合經營策略，國家到地方；地方串連到全國的計畫掌握關鍵，讓更多中小企業透過國家鼓勵獲得政策驅動力、中央計畫、地方補助、產業投資、重點扶持、輔導能量、數位科技轉型動能。

(1) 透過薪傳智庫數位科技提供 ESAA 自評表，對比更周全的國內外關聯缺口與市場開發及補助之戰情，讓組織全面系統性轉型。

(2) SESP 讓中小企業高層與利害關係人皆可快速檢討，並導出最有利商業經營策略的缺口與解決方案。

(3) 不定期提供最新的國際趨勢、商務機會、公共效益、產業串聯機會共同解決企業與社會問題價值。

3.【儲備企業 CSR 雲端數位存摺】

以經營策略新顯學來儲存疫後競爭力，提供全方位企業健檢 CSR（ESGs）以及數位資訊備查，讓臺灣政府與中小企業共好的體質持續發展獲利；薪傳聚焦深化並結合第三方公正機構聯合律師、會計師做年度 CSR 報告背書簽證，積極打造創新的價值鏈產業環境，讓政府與中小企業 CSR 存摺條條實務累積實蹟，讓每個支點帶動加乘的國際經濟效益，照亮中小企業耀享全球發展之路。

(1) 雲端數位存摺具備隨時下載提供企業自身最新 CSR 之執行實蹟與記錄，且更可隨時將前一期的評比資料（論述）調出，做為填寫新一期的參考。

(2) 隨時下載更新提供企業各部門規劃的最新 CSR 之執行實蹟與記錄。

（二）CSR 專案帶動相關效益

為企業提供各項資源與整合運用得以生存、促進五個產學共同研發專案、高雄市政府社會局網站維護、高雄市政府社會局兒少科網站建置、屏東縣財稅局政令宣導、金門酒廠全體員工幸福優退講座、活化與提高就業人口、提供在地就業機會、推動異業結盟同業聯盟、帶動地方人口創業動機、地區（域）數位群聚轉型輔導、響應推動並到各企業匯集非洲救鞋行動、每月不定期收集各企業物資分批載往各地不同單位組織、協助企業員工提升協同作業、協助企業推動員工職能與工作效益、協助企業推動員工的工作生活平衡、協助企業提升重視員工的工作環境與福利、協助企業員工提升辦公設備舒適獲得幸福感、協助企業大大降低員工流動率、協助企業的員工工作效率大幅提升、協助企業促進企業與員工誠信、協助企業員工忠誠度等工作態度、協助企業建立和凝聚員工向心力、協助企業建立品牌 LOGO 與品牌故事行銷、協助提升企業加分形象、協助企業發展國際市場、協助企業設立新南向境外公司、協助企業和平處理勞資問題、協助企業解決相關法律問題、協助企業完善財務缺口、協助企業稅務釐清與調整、協助企業疫情紓困補助、協助企業疫情紓困貸款、協助在地青年申辦青年創業、協助企業申辦微型鳳凰、協助企業辦理投資臺灣事務所三大方案、協助企業信保專案、協助企業二代接班轉型策略發展。

（三）中華薪傳智庫協會 NGO

薪傳同仁共同努力於 2020 年 3 月成立了「中華薪傳智庫協會」知識翻轉未來，薪傳書店免費課輔清寒生，辛慰的同時學生在 2020 年獲得總統教育獎，2020 年也在《今周刊》報導下、全國廣傳教育推廣知識下湧入上千萬捐款，千年之約的教育品質與消除貧窮及平等落實了個人、組織團體、地方社區、國家的點點滴滴。

薪傳最新影響力：中華薪傳智庫協會 2020 年 3 月成立，為薪傳二手書店因應疫情申辦成功藝 FUN 券獲得各地遊客來訪數，以及成功募得學生成績進步獎與獎學金募資。《今周刊》12 月大量閱讀與轉發：透視我們幫助清寒和學業低成就的學生，一對一免費課後輔導；目前每學期平均都有 170 個清寒學生，分別來自民雄、嘉義市、斗六……等八個地區。中華薪傳智庫協會湧入來自全國各地善心人士好幾百萬的捐款。綜述以上堅信薪傳智庫持續消除地區貧窮、消除學童飢餓、提升教育品質規模、照顧小朋友健康、更加重視課輔中心的飲水與環境衛生，知識讓地方就業與經濟成長脫貧，帶動城市發展永續邁向經濟命脈學無止盡之目標。

資料來源

- 王君瑭（2020 年 12 月 10 日）。最強館長棄百萬年薪！砸千萬退休金只為做「這件事」！取自 https://www.businesstoday.com.tw/article/category/183035/post/202012100011/%E6%9C%80%E5%BC%B7%E9%A4%A8%E9%95%B7%E6%A3%84%E7%99%BE%E8%90%AC%E5%B9%B4%E8%96%AA%EF%BC%81%20%E7%A0%B8%E5%8D%83%E8%90%AC%E9%80%80%E4%BC%91%E9%87%91%E9%87%91%E5%8F%AA%E7%82%BA%E5%81%9A%E3%80%8C%E9%80%99%E4%BB%B6%E4%BA%8B%E3%80%8D%EF%BC%81
- 正大聯合會計師事務所。企業社會責任報告編制。取自 https://www.grantthornton.tw/services/advisory-services/advisory-services2/
- 李介文、魏良曲（2014 年 9 月）。如何提高「企業社會責任」的效益。取自 https://www2.deloitte.com/tw/tc/pages/risk/articles/newsletter-09-24.html
- 沈瑜（2020 年 7 月 27 日）。中信金 CSR 心法：「給魚不如給釣竿」。取自 https://www.gvm.com.tw/article/73856?utm_campaign=daily&utm_medium=social&utm_source=facebook&utm_content=GV_post&fbclid=IwAR3g9O_j4lPOGsRwPtyo7yLIK587SfCenHhp189Xqe44yQtmbEOqRS5oqNY
- 科技產業資訊室（iKnow）（2006 年 6 月 1 日）。中小企業白皮書統計資料出爐。取自 https://iknow.stpi.narl.org.tw/Post/Read.aspx?PostID=857
- 高宜凡（2010 年 3 月 8 日）。企業與民眾，心中是否都長存 CSR ？取自 https://www.gvm.com.tw/article/13760
- 徐耀浤、詹世榕（2006）。從國際法觀點簡析 OECD 多國企業指導綱領之法律性質與我國之因應之思維。應用倫理研究通訊，200611（40 期），51-59。取自 http://lawdata.com.tw/tw/detail.aspx?no=180445
- 產業永續發展整合資訊網（2016 年 3 月 7 日）。企業社會責任詳細。取自 https://proj.ftis.org.tw/isdn2019/Application/Detail/6F4893445DDDBA15

三、國際指標

（一）GRI 準則為全世界 CSR 新標準

　　全球報導倡議機構 Global Reporting Initiative（GRI）於 2016 年 10 月發布了永續性報告的全球標準，為組織提供了一種公開非財務資訊的通用語言。依據 GRI 組織規定，於 2018 年 7 月 1 日起 GRI Standards.GRI 準則將取代 G4 指南，可以幫組織做出更好的決定，並為聯合國 2030 永續發展目標價值做出貢獻，成為全世界 CSR 報告的新標準。

（二）SDG18 數位科技為人類及地球謀福祉

SDG 目標 1　　消除各地一切形式的貧窮

1.1　　在西元 2030 年前，消除所有地方的極端貧窮，目前的定義為每日的生活費不到 1.25 美元。

1.1.1　低於國際貧窮線的人口比例，按性別、年齡族群、就業狀況、地理位置（城市／農村）分列。

1.2　在西元 2030 年前，依據國家的人口統計數字，將各個年齡層的貧窮男女與兒童人數減少一半。

1.2.1　低於國家貧窮線的人口比例，按性別、年齡族群分列。

1.2.2　屬於國家標準界定之各種形式貧窮的不同年齡族群男女和兒童所佔比例。

1.3　對所有的人，包括底層的人，實施適合國家的社會保護制度措施，到了西元 2030 年，範圍涵蓋貧窮與弱勢族群。

1.3.1　受到社會保障最低標準／系統涵蓋的人口比例，按性別分列，並區分為兒童、失業者、老年人、身心障礙者、孕婦、新生兒、工傷受害者、貧窮者和弱勢族群。

1.4　在西元 2030 年前，確保所有的男男女女，尤其是貧窮與弱勢族群，在經濟資源、基本服務、以及土地與其他形式的財產、繼承、天然資源、新科技與財務服務（包括微型貸款）都有公平的權利與取得權。

1.4.1　家戶中可獲得基本服務的人口比例。

1.4.2　總成年人中持有具保障的土地所有權以及法律認可文件的人口比例，按性別和所有權類型分列。

1.5　在西元 2030 年前，讓貧窮與弱勢族群具有災後復原能力，減少他們暴露於氣候極端事件與其他社經與環境災害的頻率與受傷害的嚴重度。

1.5.1　每 10 萬人中因災害死亡、失蹤和直接受影響的人數（同 11.5.1、13.1.1）。

1.5.2　災害造成的直接經濟損失與全球國內生產總值相比。

1.5.3　《2015-2030 仙台減災綱領》通過和執行國家減少災害風險策略（同 11.b.1、13.1.2）。

1.5.4　依據國家減少災害風險策略，通過和實施地方減少災害風險策略的地方政府比例（同 11.b.2、13.1.3）。

1.a　確保各個地方的資源能夠大幅動員，包括改善發展合作，為開發中國家提供妥善且可預測的方法，尤其是最低度開發國家（以下簡稱 LDCs），以實施計畫與政策，全面消除它們國內的貧窮。

1.a.1　政府直接分配給扶貧方案的資源比例。

1.a.2　花費在基本服務（包含教育、健康、社會保障）佔政府總支出的比例。

1.b　依據考量到貧窮與性別意識敏感度的發展策略，建立國家的、區域的與國際層級的妥善政策架構，加速消除貧窮行動。

1.b.1　政府經常性和資本支出中，用於婦女、貧窮、弱勢團體部門的比例。

SDG 目標 2　消除飢餓，達成糧食安全，改善營養及促進永續農業

為了迎戰貧富差距、氣候變遷、性別平權等議題，2015 年聯合國啟動「2030 永續發展目標」（Sustainable Development Goals, SDGs），提出 17 項全球政府與企業共同邁向永續發展的核心目標──第 2 項為「消除飢餓，達成糧食安全，改善營養及促進永續農業」。這項目標的條例與核心精神是什麼？國內外又有哪些實例與反思？

2.1　在西元 2030 年前，消除飢餓，確保所有的人，尤其是貧窮與弱勢族群（包括嬰兒），都能夠終年取得安全、營養且足夠的糧食。

  2.1.1　營養不良的人口比率。

  2.1.2　中度或重度糧食不安全的人口比率，以糧食不安全經驗衡量（Food Insecurity Experience Scale, FIES）為準。

2.2　西元 2030 年前，消除所有形式的營養不良，包括在西元 2025 年前，達成國際合意的 5 歲以下兒童，並且解決青少女、孕婦、哺乳婦女以及老年人的營養需求。

  2.2.1　5 歲以下兒童發育遲緩率（年齡標準身高小於世衛組織兒童生長發育標準中位數－ 2 的標準偏差）。

  2.2.2　按類型（消瘦和超重）分列的 5 歲以下兒童營養不良人口比率（身高標準體重大於或小於世衛組織兒童生長發育標準中位數＋ 2 或－ 2 的標準偏差）。

2.3　在西元 2030 年前，使農村的生產力與小規模糧食生產者的收入增加一倍，尤其是婦女、原住民、家族式農夫、牧民與漁夫，包括讓他們有安全及公平的土地、生產資源、知識、財務服務、市場、增值機會以及非農業就業機會的管道。

  2.3.1　按農業／畜牧業／林業企業規模分類的每勞動單位生產量。

  2.3.2　小規模糧食生產者的平均收入，按性別和原住民分列。

2.4　在西元 2030 年前，確保可永續發展的糧食生產系統，並實施可災後復原的農村作法，提高產能及生產力，協助維護生態系統，強化適應氣候變遷、極端氣候、乾旱、洪水與其他災害的能力，並漸進改善土地與土壤的品質。

  2.4.1　實踐具生產力和永續農業的農業面積比例。

2.5　在西元 2020 年前，維持種子、栽種植物、家畜以及與他們有關的野生品種之基因多樣性，包括善用國家、國際與區域妥善管理及多樣化的種籽與植物銀行，並確保運用基因資源與有關傳統知識所產生的好處得以依照國際協議而公平的分享。

2.5.1　保存於中長期儲存設施中之糧食和農業動植物遺傳資源之數量。

2.5.2　當地品種面臨絕種危機程度的比例（危險、沒有危險或未知）。

2.a　提高在鄉村基礎建設、農村研究、擴大服務、科技發展、植物與家畜基因銀行上的投資，包括透過更好的國際合作，以改善開發中國家的農業產能，尤其是最落後國家。

2.a.1　政府支出中農業取向政策的指數。

2.a.2　農業部門支出佔政府總支出的比例。

2.b　矯正及預防全球農業市場的交易限制與扭曲，包括依據杜哈發展圓桌，同時消除各種形式的農業出口補助及產生同樣影響的出口措施。

2.b.1　農業出口補貼。

2.c　採取措施，以確保食品與他們的衍生產品的商業市場發揮正常的功能，並如期取得市場資訊，包括儲糧，以減少極端的糧食價格波動。

2.c.1　食品價格異常指標。

SDG 目標 3　確保健康及促進各年齡層的福祉

為了迎戰貧富差距、氣候變遷、性別平權等議題，2015 年聯合國啟動「2030 永續發展目標」（Sustainable Development Goals, SDGs），提出 17 項全球政府與企業共同邁向永續發展的核心目標——第 3 項為「確保健康及促進各年齡層的福祉」。這項目標的條例與核心精神是什麼？國內外又有哪些實例與反思？

3.1　在西元 2030 年前，全球孕產婦每 10 萬例活產的死亡率降至 70 人以下。

   3.1.1　孕產婦死亡率。

   3.1.2　由專業醫護人員接生的分娩比例。

3.2　在西元 2030 年前，消除新生兒和 5 歲以下兒童可預防的死亡，各國目標將新生兒每 1,000 例活產的死亡率至少降至 12 例，5 歲以下兒童每 1,000 例活產的死亡率至少降至 25 例。

   3.2.1　5 歲以下兒童死亡率。

   3.2.2　新生兒死亡率。

3.3　在西元 2030 年前，消除愛滋病、肺結核、瘧疾以及受到忽略的熱帶性疾病，並對抗肝炎，水傳染性疾病以及其他傳染疾病。

   3.3.1　每 1,000 名未感染者中新增愛滋病毒感染病例數，按性別、年齡和關鍵族群分列。

   3.3.2　每 10 萬人中的結核病發生率。

   3.3.3　每 1,000 人中的瘧疾發生率。

   3.3.4　每 10 萬人中的 B 型肝炎發生率。

   3.3.5　需採取介入措施來治療易被輕忽的熱帶疾病的人數。

3.4　在西元 2030 年前，透過預防與治療，將非傳染性疾病的未成年死亡數減少三分之一，並促進心理健康。

   3.4.1　心血管疾病、癌症、糖尿病或慢性呼吸道疾病的死亡率。

   3.4.2　自殺死亡率。

3.5　強化物質濫用的預防與治療，包括麻醉藥品濫用以及酗酒。

   3.5.1　物質使用疾患的治療措施涵蓋範圍。

   3.5.2　根據國情定義酒精有害飲用量，以一年人均消費量計算，依 15 歲及 15 歲以上分列。

3.6　　在西元 2020 年前，讓全球因為交通事故而傷亡的人數減少一半。

　　　3.6.1　　道路交通事故死亡率。

3.7　　在西元 2030 年前，確保全球都有管道可取得性與生殖醫療保健服務，包括家庭規劃、資訊與教育，並將生殖醫療保健納入國家策略與計畫之中。

　　　3.7.1　　育齡婦女（15 至 49 歲）在計畫生育方面的需求能透過現代化方法得到滿足的比例。

　　　3.7.2　　10 至 14 歲及 15 至 19 歲年齡組，每 1,000 名女性的青少年生育率。

3.8　　實現醫療保健涵蓋全球（以下簡稱 UHC）的目標，包括財務風險保護，取得高品質基本醫療保健服務的管道，以及所有的人都可取得安全、有效、高品質、負擔得起的基本藥物與疫苗。

　　　3.8.1　　基本保健服務的涵蓋範圍。

　　　3.8.2　　與家庭支出或收入相比，家庭保健支出大的人口所佔比例。

3.9　　在西元 2030 年以前，大幅減少死於危險化學物質、空氣污染、水污染、土壤污染以及其他污染的死亡及疾病人數。

　　　3.9.1　　肇因於家戶和環境空氣污染的死亡率。

　　　3.9.2　　肇因於不安全水源與衛生設備，以及缺乏衛生環境的死亡率。

　　　3.9.3　　意外中毒死亡率。

3.a　　強化煙草管制架構公約在所有國家的實施與落實。

　　　3.a.1　　15 歲以上人口因使用煙草的患病率。

3.b　　對主要影響開發中國家的傳染以及非傳染性疾病，支援疫苗以及醫藥的研發，依據《杜哈宣言》提供負擔得起的基本藥物與疫苗；《杜哈宣言》確認開發中國家有權利使用國際專利規範 - 與貿易有關之智慧財產權協定（以下簡稱 TRIPS）中的所有供應品，以保護民眾健康，尤其是必須提供醫藥管道給所有的人。

　　　3.b.1　　國家計畫中可獲得所有疫苗的目標人口比例。

　　　3.b.2　　對醫療研究和基本衛生部門的官方發展援助總額。

　　　3.b.3　　擁有負擔得起且能永續供給的核心必要藥物的醫療設施比例。

3.c　　大幅增加開發中國家的醫療保健的融資與借款，以及醫療保健從業人員的招募、培訓以及留任，尤其是 LDCs 與小島嶼開發中國家（SIDS）。

3.c.1 　　衛生工作者密度和分布。

3.d 　強化所有國家的早期預警、風險減少，以及國家與全球健康風險的管理能力，特別是開發中國家。

3.d.1 　　國際衛生條例（IHR）的能力和衛生應急準備程度。

3.d.2 　　選擇性抗生素抗藥性有機體引發的血流感染比例。

SDG 目標 4　確保有教無類、公平以及高品質的教育，及提倡終身學習

　　為了迎戰貧富差距、氣候變遷、性別平權等議題，2015 年聯合國啟動「2030 永續發展目標」（Sustainable Development Goals, SDGs），提出 17 項全球政府與企業共同邁向永續發展的核心目標——第 4 項為「確保有教無類、公平以及高品質的教育，及提倡終身學習」。這項目標的條例與核心精神是什麼？國內外又有哪些實例與反思？

4.1 　在西元 2030 年以前，確保所有的男女學子都完成免費的、公平的以及高品質的小學與中學教育，得到有關且有效的學習成果。

4.1.1 　　兒童和青少年在 (a) 2 ／ 3 年級 、(b) 小學結束時、(c) 中學結束

時，在(i)閱讀和(ii)數學達到最低能力水平的比例，按性別分列。

4.2　在西元 2030 年以前，確保所有的孩童都能接受高品質的早期幼兒教育、照護，以及小學前教育，因而為小學的入學作好準備。

4.2.1　在健康、學習和社會心理健康方面，正常發育之 5 歲以下孩童比例，按性別分列。

4.2.2　有組織學習的參與率（正式入學年齡之前一年），按性別分列。

4.3　在西元 2030 年以前，確保所有的男女都有公平、負擔得起、高品質的技職、職業與高等教育的受教機會，包括大學。

4.3.1　過去 12 個月青年和成人正式和非正式教育和培訓的參與率，按性別分列。

4.4　在西元 2030 年以前，大幅增加掌握就業、優質工作和創業所需相關技能的青年和成年人數，包括技術與職業技能。

4.4.1　青年和成人擁有資訊傳播科技（ICT）技能的比率，按技能類型分列。

4.5　在西元 2030 年以前，消除教育上的性別不平等，確保弱勢族群有接受各階級教育的管道與職業訓練，包括身心障礙者、原住民以及弱勢孩童。

4.5.1　教育指標均等指數（所有具數據的分類統計，如女／男、鄉村／城市、財富頂端／底端之五分之一、身障狀況、原住民、受衝突影響等）。

4.6　在西元 2030 年以前，確保所有青年和大部分成年男女具有識字和算術能力。

4.6.1　特定年齡層獲得一定程度與實用的讀寫及算術能力之人口比例，按性別分列。

4.7　在西元 2030 年以前，確保所有的學子都習得必要的知識與技能而可以促進永續發展，包括永續發展教育、永續生活模式、人權、性別平等、和平及非暴力提倡、全球公民、文化差異欣賞，以及文化對永續發展的貢獻。

4.7.1　（一）全球公民教育（二）永續發展教育在 (a) 國家教育政策、(b) 課程、(c) 教師培訓、(d) 學生評估的主流化程度（同 12.8.1、13.3.1）。

4.a　建立及提升具有孩童、身心障礙以及性別意識敏感度的教育設施，並為所有的人提供安全的、非暴力的、有教無類的、以及有效的學習環境。

　　4.a.1　提供基礎服務的學校比例。

4.b　在西元 2020 年以前，大量增加全球開發中國家的獎學金數目，尤其是 LDCs、SIDS 與非洲國家，以提高高等教育的受教率，包括已開發國家與其他開發中國家的職業訓練、資訊與通信科技（以下簡稱 ICT），技術的、工程的，以及科學課程。

　　4.b.1　官方提供的獎學金數量（按部門和學習類型區分）

4.c　在西元 2030 年以前，大量增加合格師資人數，包括在開發中國家進行國際師資培訓合作，尤其是 LDCs 與 SIDS。

　　4.c.1　具有最低標準資格的教師比例（按教育程度區分）

SDG 目標 5　實現性別平等，並賦予婦女權力

　　為了迎戰貧富差距、氣候變遷、性別平權等議題，2015 年聯合國啟動「2030 永續發展目標」（Sustainable Development Goals, SDGs），提出 17 項全

球政府與企業共同邁向永續發展的核心目標──第 5 項為「實現性別平等，並賦予婦女權力」。這項目標的條例與核心精神是什麼？國內外又有哪些實例與反思？

5.1　消除全球對婦女和女童一切形式的歧視。

　　5.1.1　制定法律框架來促進、推行和監督實現性別平等和無歧視。

5.2　消除公私領域針對婦女和女童一切形式的暴力，包括販運、性剝削及其他形式的剝削。

　　5.2.1　有過伴侶的婦女和 15 歲及以上女童，在過去 12 個月中遭受現任或前任伴侶毆打、性暴力或心理暴力的比例，按暴力形式和年齡分列。

　　5.2.2　婦女和 15 歲及以上女童，在過去 12 個月中遭受親密伴侶以外之其他人的性暴力比例，按年齡和發生地分列。

5.3　消除童婚、早婚、強迫婚姻及女性割禮等一切傷害行為。

　　5.3.1　20 至 24 歲婦女，在 15 歲和 18 歲之前結婚或同居的婦女所佔比例。

　　5.3.2　15 至 49 歲女童和婦女，生殖器被殘割／切割的人所佔比例，按年齡分列。

5.4　認可和尊重無償的照護和家務工作，國家依國情提供公共服務、基礎設施和社會保護政策，並提倡家庭成員共同分擔責任。

　　5.4.1　用於無薪酬家務和家庭照護工作的時間所佔比例，按性別、年齡和地點分列。

5.5　確保婦女全面有效參與各級政經和公共決策，並享有參與各級決策領導層級的平等機會。

　　5.5.1　婦女在國家議會和地方政府席位中所佔比例。

　　5.5.2　婦女擔任管理職務的比例。

5.6　依據國際人口與發展會議（ICPD）行動計畫、北京行動平台，以及它們的檢討成果書，確保每個地方的人都有管道取得性與生殖醫療照護服務。

　　5.6.1　15 至 49 歲婦女就性關係、避孕措施和生育保健議題，自主做出知情決定的比例。

5.6.2 國家制定法律規章確保 15 歲及以上的男女，充分和平等享有獲取性和生育保健的資訊和教育機會。

5.a 進行改革，以提供婦女公平的經濟資源權利，以及土地與其他形式的財產、財務服務、繼承與天然資源的所有權與掌控權。

5.a.1 (a) 擁有農地所有權或保障權的男女性別佔農業總人口的比例；(b) 女性佔農地所有權人或權利人的比例。

5.a.2 國家法律保障女性對於土地所有權或其他控制權擁有平等權利的比例。

5.b 改善科技的使用能力，特別是 ICT，以提高婦女的能力。

5.b.1 擁有行動電話的個人比例（按性別區分）。

5.c 採用及強化完善的政策以及可實行的立法，以促進性別平等，並提高各個階層婦女的能力。

5.c.1 國家有系統地追蹤和分配公共撥款用於性別平等和賦予婦女權力的比例。

SDG 目標 6 確保所有人都能享有水及衛生及其永續管理

　　為了迎戰貧富差距、氣候變遷、性別平權等議題，2015 年聯合國啟動「2030 永續發展目標」（Sustainable Development Goals, SDGs），提出 17 項全球政府與企業共同邁向永續發展的核心目標──第 6 項為「確保所有人都能享有水及衛生及其永續管理」。這項目標的條例與核心精神是什麼？國內外又有哪些實例與反思？

6.1　在西元 2030 年以前，讓全球的每一個人都有公平的管道，可以取得安全且負擔得起的飲用水。

　　6.1.1　使用得到安全管理飲用水服務的人口比例。

6.2　在西元 2030 年以前，讓每一個人都享有公平及妥善的衛生，終結露天大小便，特別注意弱勢族群中婦女的需求。

　　6.2.1　使用得到安全管理的環境衛生設施服務（包括提供肥皂和水的洗手設施）的人口比例。

6.3　在西元 2030 年以前，改善水質，減少污染，消除垃圾傾倒，減少有毒物化學物質與危險材料的釋出，將未經處理的廢水比例減少一半，大量提高全球的回收與安全再使用率。

　　6.3.1　安全處理廢水的比例。

　　6.3.2　環境水質良好的水體比例。

6.4　在西元 2030 年以前，大幅增加各個產業的水使用效率，確保永續的淡水供應與回收，以解決水饑荒問題，並大幅減少因為水計畫而受苦的人數。

　　6.4.1　按時間列出的用水效率變化。

　　6.4.2　用水壓力：淡水抽取量佔可用淡水資源的比例。

6.5　在西元 2030 年以前，全面實施一體化的水資源管理，包括跨界合作。

　　6.5.1　水資源整合管理執行程度。

　　6.5.2　制定有水資源合作計畫的跨界流域比例。

6.6　在西元 2020 年以前，保護及恢復跟水有關的生態系統，包括山脈、森林、沼澤、河流、含水層，以及湖泊。

　　6.6.1　與水有關的生態系統範圍隨時間的變化。

6.a　在西元 2030 年以前，針對開發中國家的水與衛生有關活動與計畫，擴大國際合作與能力培養支援，包括採水、去鹽、水效率、廢水處理、回收，以及再使用科技。

6.a.1　在政府預算中，與水及衛生相關的政府發展援助金額。

6.b　支援及強化地方社區的參與，以改善水與衛生的管理。

6.b.1　具有地方社區參與水和衛生管理的既定和業務政策和程序的地方行政單位的比例。

SDG 目標 7　確保所有的人都可取得負擔得起、可靠的、永續的，及現代的能源

為了迎戰貧富差距、氣候變遷、性別平權等議題，2015 年聯合國啟動「2030 永續發展目標」（Sustainable Development Goals, SDGs），提出 17 項全球政府與企業共同邁向永續發展的核心目標——第 7 項為「確保所有的人都可取得負擔得起、可靠的、永續的，及現代的能源」。這項目標的條例與核心精神是什麼？國內外又有哪些實例與反思？

7.1　在西元 2030 年前，確保所有的人都可取得負擔得起、可靠的，以及現代的能源服務。

7.1.1　獲得供電的人口比例。

7.1.2　主要依靠清潔燃料和技術的人口比例。

7.2　在西元 2030 年以前，大幅提高全球再生能源的共享。

7.2.1　再生能源在最終能源消費總量中的份額。

7.3　在西元 2030 年以前，將全球能源效率的改善率提高一倍。

7.3.1　以初級能源和國內生產總值計算的能源密集度。

7.a　在西元 2030 年以前，改善國際合作，以提高乾淨能源與科技的取得管道，包括再生能源、能源效率、更先進及更乾淨的石化燃料科技，並促進能源基礎建設與乾淨能源科技的投資。

7.a.1　國際資金流動至開發中國家以支持乾淨能源的研究發展與製造可再生能源，包括混合系統。

7.b　在西元 2030 年以前，擴大基礎建設並改善科技，以為所有開發中國家提供現代及永續的能源服務，尤其是 LDCs 與 SIDS。

7.b.1　開發中國家製造可再生能源的能力（同 12.a.1）。

SDG 目標 8　促進包容且永續的經濟成長，達到全面且有生產力的就業，讓每一個人都有一份好工作

為了迎戰貧富差距、氣候變遷、性別平權等議題，2015 年聯合國啟動「2030 永續發展目標」（Sustainable Development Goals, SDGs），提出 17 項全球政府與企業共同邁向永續發展的核心目標——第 8 項為「促進包容且永續的經濟成長，達到全面且有生產力的就業，讓每一個人都有一份好工作」。這項目標的條例與核心精神是什麼？國內外又有哪些實例與反思？

8.1 依據國情維持經濟成長，尤其是開發度最低的國家，每年的國內生產毛額（以下簡稱 GDP）成長率至少 7%。

    8.1.1 實質人均國內生產總值年增長率。

8.2 透過多元化、科技升級與創新提高經濟體的產能，包括將焦點集中在高附加價值與勞動力密集的產業。

    8.2.1 就業人員實質人均國內生產總值年增長率。

8.3 促進以開發為導向的政策，支援生產活動、就業創造、企業管理、創意與創新，並鼓勵微型與中小企業的正式化與成長，包括取得財務服務的管道。

    8.3.1 非正式就業在就業機會中的比例，按性別分列。

8.4 在西元 2030 年以前，漸進改善全球的能源使用與生產效率，在已開發國家的帶領下，依據十年的永續使用與生產計畫架構，努力減少經濟成長與環境惡化之間的關聯。

    8.4.1 物質足跡、人均物質足跡和單位國內生產總值的物質足跡（同 12.2.1）。

    8.4.2 國內物質消費、人均國內物質消費和單位國內生產總值的國內物質消費（同 12.2.2）。

8.5 在西元 2030 年以前，實現全面有生產力的就業，讓所有的男女都有一份好工作，包括年輕人與身心障礙者，並實現同工同酬的待遇。

    8.5.1 員工的平均時薪，按性別、職業、年齡、身心障礙者分列。

    8.5.2 失業率，按性別、年齡、身心障礙者分列。

8.6 在西元 2020 年以前，大幅減少失業、失學或未接受訓練的年輕人。

    8.6.1 15 至 24 歲青年，未受教育、就業或培訓之人數比例。

8.7 採取立即且有效的措施，以禁止與消除最糟形式的童工，消除受壓迫的勞工；在西元 2025 年以前，終結各種形式的童工，包括童兵的招募使用。

    8.7.1 5 至 17 歲兒童從事童工勞動的比例和人數，按性別和年齡分列。

8.8 保護勞工的權益，促進工作環境的安全，包括遷徙性勞工，尤其是婦女以及實行危險工作的勞工。

    8.8.1 致死和非致死的職業傷害頻率，按性別和移民身分分列。

    8.8.2 遵守國際勞工組織和國家法律規範的勞工權利之程度，包含結社自由和集體談判，按性別和移民身分分列。

8.9 在西元 2030 年以前，制定及實施政策，以促進永續發展的觀光業，創造就業，促進地方文化與產品。

    8.9.1 旅遊業國內生產總值佔總額之比例和成長率。

    8.9.2 永續旅遊業工作數量佔旅遊業工作總數之比例。

8.10 強化本國金融機構的能力，為所有的人提供更寬廣的銀行、保險與金融服務。

    8.10.1 每 10 萬成年人可使用的商業銀行分行數目可使用的自動提款機（ATM）數目。

    8.10.2 15 歲及以上成年人在銀行、其他金融機構或手機行動支付擁有帳戶之比例。

8.a 提高給開發中國家的貿易協助資源，尤其是 LDCs，包括為 LDCs 提供更好的整合架構。

    8.a.1 援助貿易承諾和支付。

8.b 在西元 2020 年以前，制定及實施年輕人就業全球策略，並落實全球勞工組織的全球就業協定。

    8.b.1 政府制定已發展完成且運作中的青年就業國家政策，作為獨立政策或併入國家就業政策。

> **SDG 目標 9** 建立具有韌性的基礎建設，促進包容且永續的工業，並加速創新

為了迎戰貧富差距、氣候變遷、性別平權等議題，2015 年聯合國啟動「2030 永續發展目標」（Sustainable Development Goals, SDGs），提出 17 項全球政府與企業共同邁向永續發展的核心目標——第 9 項為「建立具有韌性的基礎建設，促進包容且永續的工業，並加速創新」。這項目標的條例與核心精神是什麼？國內外又有哪些實例與反思？

9.1　發展高品質的、可靠的、永續的，以及具有災後復原能力的基礎設施，包括區域以及跨界基礎設施，以支援經濟發展和人類福祉，並將焦點放在為所有的人提供負擔得起又公平的管道。

　9.1.1　居住在四季通行的道路 2 公里之內的農村人口所佔比例。

　9.1.2　客運量和貨運量，按運輸方式分列。

9.2　促進包容以及永續的工業化，在西元 2030 年以前，依照各國的情況大幅提高工業的就業率與 GDP，尤其是 LDCs 應增加一倍。

　9.2.1　製造業附加值佔國內生產總值的比例和人均值。

9.2.2 製造業的就業佔總就業的比例。

9.3 提高小規模工商業取得金融服務的管道，尤其是開發中國家，包括負擔得起的貸款，並將他們併入價值鏈與市場之中。

9.3.1 小型工業在工業總附加值中之比例。

9.3.2 小型工業中獲得貸款或信貸額度的比例。

9.4 在西元 2030 年以前，升級基礎設施，改造工商業，使他們可永續發展，提高能源使用效率，大幅採用乾淨又環保的科技與工業製程，所有的國家都應依據他們各自的能力行動。

9.4.1 每單位附加值的二氧化碳排放量。

9.5 改善科學研究，提高五所有國家的工商業的科技能力，尤其是開發中國家，包括在西元 2030 年以前，鼓勵創新，大量提高每百萬人中的研發人員數，並提高公民營的研發支出。

9.5.1 研發支出佔國內生產總值的比例。

9.5.2 每百萬居民中的全職研究員人數。

9.a 透過改善給非洲國家、LDCs、內陸開發中國家（以下簡稱 LLDCs）與 SIDS 的財務、科技與技術支援，加速開發中國家發展具有災後復原能力且永續的基礎設施。

9.a.1 政府對於基礎設施所展現的國際支持程度。

9.b 支援開發中國家的本國科技研發與創新，包括打造有助工商多元發展以及商品附加價值提升的政策環境。

9.b.1 中高新技術產業增值佔總增值的比重。

9.c 大幅提高 ICT 的管道，在西元 2020 年以前，在開發度最低的發展中國家致力提供人人都可取得且負擔得起的網際網路管道。

9.c.1 行動網路覆蓋的人口比例。

SDG 目標 10　減少國內及國家間不平等

　　為了迎戰貧富差距、氣候變遷、性別平權等議題，2015 年聯合國啟動「2030 永續發展目標」（Sustainable Development Goals, SDGs），提出 17 項全球政府與企業共同邁向永續發展的核心目標——第 10 項為「減少國內及國家間不平等」。這項目標的條例與核心精神是什麼？國內外又有哪些實例與反思？

10.1　在西元 2030 年以前，以高於國家平均值的速率漸進地致使底層 40% 的人口實現所得成長。

　　10.1.1　最底層 40% 人口和總人口的家庭支出或人均所得成長率。

10.2　在西元 2030 年以前，促進社經政治的融合，無論年齡、性別、身心障礙、種族、人種、祖國、宗教、經濟或其他身分地位。

　　10.2.1　收入低於收入中位數 50% 的人口所佔比例，依性別、年齡和身心障礙者分列。

10.3　確保機會平等，減少不平等，作法包括消除歧視的法律、政策及實務作法，並促進適當的立法、政策與行動。

10.3.1 　根據《國際人權法》所禁止的歧視理由，過去 12 個月內個人受到歧視或騷擾的人口報告比例（同 16.b.1）。

10.4 　採用適當的政策，尤其是財政、薪資與社會保護政策，並漸進實現進一步的平等。

10.4.1 　勞動力佔國內生產總值的份額，包括工資和社會保障轉移。

10.5 　改善全球金融市場與金融機構的法規與監管，並強化這類法規的實施。

10.5.1 　金融健全性指標。

10.6 　提高發展中國家在全球經濟與金融機構中的決策發言權，以實現更有效、更可靠、更負責以及更正當的機構。

10.6.1 　開發中國家在國際組織中的成員和表決權的比例（同 16.8.1）。

10.7 　促進有秩序的、安全的、規律的，以及負責的移民，作法包括實施規劃及管理良好的移民政策。

10.7.1 　由員工負擔的招募費用佔其在當地年收入的比例。

10.7.2 　國家實施管理完善的移民政策。

10.a 　依據世界貿易組織（以下簡稱 WTO）的協定，對開發中國家實施特別且差異對待的原則，尤其是開發度最低的國家。

10.a.1 　LDCs 和零關稅發展中國家進口的關稅線的比例。

10.b 　依據國家計畫與方案，鼓勵官方開發援助（ODA）與資金流向最需要的國家，包括外資直接投資，尤其是 LDCs、非洲國家、SIDS，以及 LLDCs。

10.b.1 　受援國和捐助國的資源流動總額和流動類型（如 ODA、外資直接投資和其他流動）。

10.c 　在西元 2030 年以前，將遷移者的匯款手續費減少到小於 3%，並消除手續費高於 5% 的匯款。

10.c.1 　匯款成本佔匯款金額的比例。

SDG 目標 11　促使城市與人類居住具包容、安全、韌性及永續性

　　為了迎戰貧富差距、氣候變遷、性別平權等議題，2015 年聯合國啟動「2030 永續發展目標」(Sustainable Development Goals, SDGs)，提出 17 項全球政府與企業共同邁向永續發展的核心目標——第 11 項為「促使城市與人類居住具包容、安全、韌性及永續性」。這項目標的條例與核心精神是什麼？國內外又有哪些實例與反思？

11.1　在西元 2030 年前，確保所有的人都可取得適當的、安全的，以及負擔得起的住宅與基本服務，並改善貧民窟。

　　11.1.1　居住在貧民窟、非正規居住區或住房不足之城市人口比例。

11.2　在西元 2030 年以前，為所有的人提供安全的、負擔得起、可使用的，以及可永續發展的交通運輸系統，改善道路安全，尤其是擴大公共運輸，特別注意弱勢族群、婦女、兒童、身心障礙者以及老年人的需求。

　　11.2.1　擁有便捷之公共運輸的人口比例，按性別、年齡和身心障礙者分列。

11.3　在西元 2030 年以前，提高融合的、包容的以及可永續發展的都市化與容積，以讓所有的國家落實參與性、一體性以及可永續發展的人類定居

規劃與管理。

　11.3.1　土地使用率與人口成長率之比率。

　11.3.2　民間社會透過以民主方式定期運作的架構，直接參與城市規劃與管理的城市所佔百分比。

11.4　在全球的文化與自然遺產的保護上，進一步努力。

　11.4.1　用於保存、保護和養護所有文化與自然遺產的人均支出總額，按遺產類型，政府層級，支出類型和私人資金類型列出。

11.5　在西元 2030 年以前，大幅減少災害的死亡數以及受影響的人數，並大幅減少災害所造成的 GDP 經濟損失，包括跟水有關的傷害，並將焦點放在保護弱勢族群與貧窮者。

　11.5.1　每 10 萬人中因災害死亡、失蹤及直接受影響的人數（同 1.5.1、13.1.1）。

　11.5.2　災害造成與全球國內生產總值相關的直接經濟損失、重要基礎設施損壞和基本服務中斷次數。

11.6　在西元 2030 年以前，減少都市對環境的有害影響，其中包括特別注意空氣品質、都市管理與廢棄物管理。

　11.6.1　定期收集並得到適當最終處理的城市固體廢棄物，佔城市固體廢棄物總量之比例，按城市列出。

　11.6.2　按人口權重計算的城市 顆粒物（例如 $PM_{2.5}$ 和 PM_{10}）年度均值。

11.7　在西元 2030 年以前，為所有的人提供安全的、包容的、可使用的綠色公共空間，尤其是婦女、孩童、老年人以及身心障礙者。

　11.7.1　城市建設區中供所有人使用的開放公共空間的平均比例，按性別、年齡和身心障礙者分列。

　11.7.2　過去 12 個月 8 中受到身體騷擾或性騷擾之受害者比例，按性別、年齡、身心障礙者和發生地點分列。

11.a　強化國家與區域的發展規劃，促進都市、郊區與城鄉之間的社經與環境的正面連結。

　11.a.1　具有 (a) 回應人口動態、(b) 確保地域發展平衡、(c) 增加財政空間的國家城市政策或地區發展計畫的國家數量

11.b 在西元 2020 年以前，大量增加在包容、融合、資源效率、移民、氣候變遷適應、災後復原能力上落實一體政策與計畫的都市與地點數目，依照日本兵庫縣架構管理所有階層的災害風險。

　11.b.1 依據《2015-2030 仙台減災綱領》通過和執行國家減少災害風險策略（同 1.5.3、13.1.2）。

　11.b.2 依據國家減少災害風險策略，通過和實施地方減少災害風險策略的地方政府比例（同 1.5.4、13.1.3）。

11.c 支援開發度最低的國家，以妥善使用當地的建材，營建具有災後復原能力且可永續的建築，作法包括財務與技術上的協助。

SDG 目標 12 ｜ 確保永續消費及生產模式

　為了迎戰貧富差距、氣候變遷、性別平權等議題，2015 年聯合國啟動「2030 永續發展目標」（Sustainable Development Goals, SDGs），提出 17 項全球政府與企業共同邁向永續發展的核心目標——第 12 項為「保永續消費及生產模式」。這項目標的條例與核心精神是什麼？國內外又有哪些實例與反思？

12.1 實施永續消費與生產十年計畫架構（以下簡稱 10YEP），所有的國家動起來，由已開發國家擔任帶頭角色，考量開發中國家的發展與能力。

12.1.1 制定永續消費和生產的國家行動計畫，或已將永續消費和生產列為國家政策主軸的優先事項或目標。

12.2 在西元 2030 年以前，實現自然資源的永續管理以及有效率的使用。

12.2.1 物質足跡、人均物質足跡和單位國內生產總值的物質足跡（同 8.4.1）。

12.2.2 國內物質消費、人均國內物質消費和單位國內生產總值的國內物質消費（同 8.4.2）。

12.3 在西元 2030 年以前，將零售與消費者階層上的全球糧食浪費減少一半，並減少生產與供應鏈上的糧食損失，包括採收後的損失。

12.3.1 全球糧食損耗指數。

12.4 在西元 2020 年以前，依據議定的國際架構，在化學藥品與廢棄物的生命週期中，以符合環保的方式妥善管理化學藥品與廢棄物，大幅減少他們釋放到空氣、水與土壤中，以減少他們對人類健康與環境的不利影響。

12.4.1 根據有害廢棄物及其他化學品的國際多邊環境協定，按照每個相關協定的要求履行資訊傳送承諾和義務。

12.4.2 有害廢棄物人均生成量，按處理方式之比例分列。

12.5 在西元 2030 年以前，透過預防、減量、回收與再使用大幅減少廢棄物的產生。

12.5.1 國家回收利用率、物資回收利用噸數。

12.6 鼓勵企業採取可永續發展的工商作法，尤其是大規模與跨國公司，並將永續性資訊納入他們的報告週期中。

12.6.1 發布永續性報告的企業數量。

12.7 依據國家政策與優先要務，促進可永續發展的公共採購流程。

12.7.1 實施永續公共採購政策和行動計畫。

12.8 在西元 2030 年以前，確保每個地方的人都有永續發展的有關資訊與意識，以及跟大自然和諧共處的生活方式。

12.8.1 （一）全球公民教育（二）永續發展教育在 (a) 國家教育政策、

(b) 課程、(c) 教師培訓、(d) 學生評估的主流化程度（同 4.7.1、13.3.1）。

12.a 協助開發中國家強健它們的科學與科技能力，朝向更能永續發展的耗用與生產模式。

    12.a.1 開發中國家製造可再生能源的能力（同 7.b.1）。

12.b 制定及實施政策，以監測永續發展對創造就業，促進地方文化與產品的永續觀光的影響。

    12.b.1 實施標準會計工具以監控經濟與環境方面的旅遊業永續發展。

12.c 依據國情消除市場扭曲，改革鼓勵浪費的無效率石化燃料補助，作法包括改變課稅架構，逐步廢除這些有害的補助，以反映他們對環境的影響，全盤思考開發中國家的需求與狀況，以可以保護貧窮與受影響社區的方式減少它們對發展的可能影響。

    12.c.1 化石燃料補助數量。

SDG 目標 13　採取緊急措施以因應氣候變遷及其影響

　　為了迎戰貧富差距、氣候變遷、性別平權等議題，2015 年聯合國啟動「2030 永續發展目標」（Sustainable Development Goals, SDGs），提出 17 項全球政府與企業共同邁向永續發展的核心目標──第 13 項為「採取緊急措施以因應氣候變遷及其影響」。這項目標的條例與核心精神是什麼？國內外又有哪些實例與反思？

13.1　強化所有國家對天災與氣候有關風險的災後復原能力與調適適應能力。

　　13.1.1　每 10 萬人中因災害死亡、失蹤和直接受影響的人數（同 1.5.1、11.5.1）。

　　13.1.2　依據《2015-2030 仙台減災綱領》通過和執行國家減少災害風險策略（同 1.5.3、11.b.1）。

　　13.1.3　依據國家減少災害風險策略，通過和實施地方減少災害風險策略的地方政府比例（同 1.5.4、11.b.2）。

13.2　將氣候變遷措施納入國家政策、策略與規劃之中。

　　13.2.1　建立或實施整合政策／策略／計畫，以增加適應氣候變遷不利影響的能力，並以不影響糧食生產的方式，促進氣候抗禦能力和低溫室氣體排放的發展（包括國家適應計畫、國家自定預期貢獻、國家通訊傳播、兩年期更新報告等）的國家數量。

13.3　在氣候變遷的減險、適應、影響減少與早期預警上，改善教育，提升意識，增進人與機構的能力。

　　13.3.1　（一）全球公民教育（二）永續發展教育在 (a) 國家教育政策、(b) 課程、(c) 教師培訓、(d) 學生評估的主流化程度（同 4.7.1、12.8.1）。

　　13.3.2　國家為執行調適、減緩、技術轉讓和發展行動，而加強機構、系統和個人能力建設。

13.a　在西元 2020 年以前，落實 UNFCCC 已開發國家簽約國的承諾，目標是每年從各個來源募得美元 1 千億，以有意義的減災與透明方式解決開發中國家的需求，並盡快讓綠色氣候基金透過資本化而全盤進入運作。

　　13.a.1　根據現存的募資 1 千億美元目標承諾，每年募得的資金（以美元計）。

13.b 提升開發度最低國家中的有關機制，以提高能力而進行有效的氣候變遷規劃與管理，包括將焦點放在婦女、年輕人、地方社區與邊緣化社區。

    13.b.1 建立或實施整合政策／策略／計畫，以增加適應氣候變遷不利影響的能力，並以不影響糧食生產的方式，促進氣候抗禦能力和低溫室氣體排放的發展（包括國家適應計劃、國家自定預期貢獻、國家通訊傳播、兩年期更新報告等）的 LDCs 與 SIDS 數量。

SDG 目標 14   保育及永續利用海洋與海洋資源，以確保永續發展

14.1 在西元 2025 年以前，預防及大幅減少各式各樣的海洋污染，尤其是來自陸上活動的污染，包括海洋廢棄物以及營養污染。

    14.1.1 沿岸優養化指數（IECP）和漂浮塑膠碎片密度。

14.2 在西元 2020 年以前，以可永續的方式管理及保護海洋與海岸生態，避免重大的不利影響，作法包括強健他們的災後復原能力，並採取復原動作，以實現健康又具有生產力的海洋。

14.2.1　國家經濟特區中以生態系統管理措施為基礎進行管理的比例。

14.3　減少並解決海洋酸化的影響，作法包括改善所有階層的科學合作。

14.3.1　具共識的一系列具代表性的採樣站所測量之平均海洋酸度（pH值）。

14.4　在西元 2020 年以前，有效監管採收，消除過度漁撈，以及非法的、未報告的、未受監管的（IUU）、或毀滅性魚撈作法，並實施科學管理計畫，在最短的時間內，將魚量恢復到依據它們的生物特性可產生最大永續發展的魚量。

14.4.1　在生物永續產量水平範圍內的魚群比例。

14.5　在西元 2020 年以前，依照國家與國際法規，以及可取得的最佳科學資訊，保護至少 10% 的海岸與海洋區。

14.5.1　保護區面積佔海洋區域的比例。

14.6　在西元 2020 年以前，禁止會造成過度魚撈的補助，消除會助長 IUU 魚撈的補助，禁止引入這類補助，承認對開發中國家與開發度最低國家採取適當且有效的特別與差別待遇應是 WTO 漁撈補助協定的一部分。

14.6.1　國家依據國際文件執行打擊非法、未報告和無管制（IUU）捕撈活動的進展。

14.7　在西元 2030 年以前，提高海洋資源永續使用對 SIDS 與 LDCs 的經濟好處，作法包括永續管理漁撈業、水產養殖業與觀光業。

14.7.1　SIDS、LDCs 和所有國家的永續漁業佔國內生產總值的比例。

14.a　提高科學知識，發展研究能力，轉移海洋科技，思考跨政府海洋委員會的海洋科技轉移準則，以改善海洋的健康，促進海洋生物多樣性對開發中國家的發展貢獻，特別是 SIDS 與 LDCs。

14.a.1　總研究預算當中，分配至海洋科技領域的比例。

14.b　提供小規模人工魚撈業者取得海洋資源與進入市場的管道。

14.b.1　保護小規模漁業權利之法規、政策、措施的實施程度。

14.c 確保《聯合國海洋法公約》（以下簡稱 UNCCLOS）簽約國全面落實國際法，包括現有的區域與國際制度，以保護及永續使用海洋及海洋資源。

14.c.1 藉由立法、政策、制度架構、海洋相關文件等方式落實國際法，回應聯合國海洋法公約》，成為保護及永續利用海洋資源的國家數量。

SDG 目標 15 保護、維護及促進領地生態系統的永續使用，永續的管理森林，對抗沙漠化，終止及逆轉土地劣化，並遏止生物多樣性的喪失

為了迎戰貧富差距、氣候變遷、性別平權等議題，2015 年聯合國啟動「2030 永續發展目標」（Sustainable Development Goals, SDGs），提出 17 項全球政府與企業共同邁向永續發展的核心目標——第 15 項為「保護、維護及促進領地生態系統的永續使用，永續的管理森林，對抗沙漠化，終止及逆轉土地劣化，並遏止生物多樣性的喪失」。這項目標的條例與核心精神是什麼？國內外又有哪些實例與反思？

15.1 在西元 2020 年以前，依照在國際協定下的義務，保護、恢復及永續使用領地與內陸淡水生態系統與他們的服務，尤其是森林、沼澤、山脈與旱地。

15.1.1 森林面積佔土地總面積的比例。

15.1.2 陸地及淡水生物多樣性的重要場域有被納入保護區的比例，按生態系統類型分列。

15.2 在西元 2020 年以前，進一步落實各式森林的永續管理，終止毀林，恢復遭到破壞的森林，並大量增加全球的造林。

15.2.1 實施永續森林管理的進展。

15.3 在西元 2020 年以前，對抗沙漠化，恢復惡化的土地與土壤，包括受到沙漠化、乾旱及洪水影響的地區，致力實現沒有土地破壞的世界。

15.3.1 已退化土地面積佔土地總面積的比例。

15.4 在西元 2030 年以前，落實山脈生態系統的保護，包括他們的生物多樣性，以改善他們提供有關永續發展的有益能力。

15.4.1 山區生物多樣性之重要場域有被納入保護區的面積。

15.4.2 山區綠化覆蓋指數。

15.5 採取緊急且重要的行動減少自然棲息地的破壞，終止生物多樣性的喪失，在西元 2020 年以前，保護及預防瀕危物種的絕種。

15.5.1 紅皮書指數。

15.6 確保基因資源使用所產生的好處得到公平公正的分享，促進基因資源使用的適當管道。

15.6.1 國家通過立法、行政和政策框架以確保公正和公平分享利益。

15.7 採取緊急動作終止受保護動植物遭到盜採、盜獵與非法走私，並解決非法野生生物產品的供需。

15.7.1 被盜獵或非法販賣的野生動物的比例（同 15.c.1）。

15.8 在西元 2020 年以前，採取措施以避免侵入型外來物種入侵陸地與水生態系統，且應大幅減少他們的影響，並控管或消除優種。

15.8.1 國家通過國家立法和投入充分資源以預防或控制外來入侵物種。

15.9 在西元 2020 年以前，將生態系統與生物多樣性價值納入國家與地方規劃、發展流程與脫貧策略中。

15.9.1 根據「2011-2020 年生物多樣性策略計畫」，在「愛知生物多樣性目標 2」的國家目標之進展。

15.a 動員並大幅增加來自各個地方的財物資源，以保護及永續使用生物多樣性與生態系統。

15.a.1 關於生物多樣性和生態系統保護和永續利用的 ODA 和公共支出（同 15.b.1）。

15.b 大幅動員來自各個地方的各階層的資源，以用於永續森林管理，並提供適當的獎勵給開發中國家改善永續森林管理，包括保護及造林。

15.b.1 關於生物多樣性和生態系統保護和永續利用的 ODA 和公共支出（同 15.a.1）。

15.c 改善全球資源，以對抗保護物種的盜採、盜獵與走私，作法包括提高地方社區的能力，以追求永續發展的謀生機會。

15.c.1 被盜獵或非法販賣的野生動物的比例（同 15.7.1）。

| SDG 目標 16 | 促進和平且包容的社會，以落實永續發展；提供司法管道給所有人；在所有階層建立有效的、負責的且包容的制度 |

　　為了迎戰貧富差距、氣候變遷、性別平權等議題，2015 年聯合國啟動「2030 永續發展目標」（Sustainable Development Goals, SDGs），提出 17 項全球政府與企業共同邁向永續發展的核心目標——第 16 項為「促進和平且包容的社會，以落實永續發展；提供司法管道給所有人；在所有階層建立有效的、負責的且包容的制度」。這項目標的條例與核心精神是什麼？國內外又有哪些實例與反思？

16.1　大幅減少各地各種形式的暴力以及有關的死亡率。

　　16.1.1　每 10 萬人中故意殺人案的受害者人數，依性別和年齡分列。

　　16.1.2　每 10 萬人中與衝突有關的死亡人數，依性別、年齡和死因分列。

　　16.1.3　過去 12 個月內遭受身心和性暴力侵害的人口比例。

　　16.1.4　在居住區單獨步行感到安全的人口比例。

16.2　終結各種形式的兒童虐待、剝削、走私、暴力以及施虐。

　　16.2.1　過去一個月內受到照顧者施加的任何體罰和心理侵害的 1 至 17

歲兒童比例。

16.2.2 10 萬人中受到人口販運行為受害者的人數，依性別、年齡和剝削形式分列。

16.2.3 18 歲之前受到性暴力侵害的 18 至 29 歲青年男女的比例。

16.3 促進國家與國際的法則，確保每個人都有公平的司法管道。

16.3.1 過去 12 個月內，向主管機關或其他官方認可的衝突解決機構通報其受害經歷的暴力行為受害者比例。

16.3.2 未判刑的被拘留者佔監獄服刑總人口的比例。

16.4 在西元 2030 年以前，大幅減少非法的金錢與軍火流，提高失物的追回，並對抗各種形式的組織犯罪。

16.4.1 流入和流出的非法資金流量總值，以美金現值計。

16.4.2 其非法來源或背景已被主管當局按照國際文書規定追查或確定的已繳獲、發現或交出的武器比例。

16.5 大幅減少各種形式的貪污賄賂。

16.5.1 過去 12 個月內至少與公職人員接觸過一次，向公職人員行賄或被公職人員要求行賄的人所佔比例。

16.5.2 過去 12 個月內至少與公職人員接觸過一次，向公職人員行賄或被公職人員要求行賄的企業所佔比例。

16.6 在所有的階層發展有效的、負責的且透明的制度。

16.6.1 政府基本支出佔原核定預算的比例，依部門分列。

16.6.2 對前次公共服務經驗感到滿意的人口比例。

16.7 確保各個階層的決策回應民意，是包容的、參與的且具有代表性。

16.7.1 公共機構中的職位相對全國分配數的比例，依性別、年齡、身心障礙者和人口群體分列。

16.7.2 認為決策具包容性和回應性的人口比例，依性別、年齡、身心障礙者、人口群體分列。

16.8 擴大及強化開發中國家參與全球管理制度。

16.8.1 開發中國家在國際組織中的成員和投票權的比例（同 10.6.1）。

16.9 在西元 2030 年以前，為所有的人提供合法的身分，包括出生登記。

16.9.1 已進行出生登記的 5 歲以下兒童比例，按年齡分列。

16.10 依據國家立法與國際協定，確保民眾可取得資訊，並保護基本自由。

16.10.1 過去 12 個月內記者、相關媒體人員、工會會員和人權倡導者被殺害、綁架、強迫失蹤、任意拘留和施以酷刑的經核實案件數目。

16.10.2 國家通過並執行憲法、法律和政策，以保障民眾取得資訊。

16.a 強化有關國家制度，作法包括透過國際合作，以建立在各個階層的能力，尤其是開發中國家，以預防暴力並對抗恐怖主義與犯罪。

16.a.1 擁有符合《巴黎原則》的獨立的國家人權機構。

16.b 促進及落實沒有歧視的法律與政策，以實現永續發展。

16.b.1 根據《國際人權法》所禁止的歧視理由，過去 12 個月內個人受到歧視或騷擾的人口報告比例（同 10.3.1）。

SDG 目標 17　強化永續發展執行方法及活化永續發展全球夥伴關係

為了迎戰貧富差距、氣候變遷、性別平權等議題，2015 年聯合國啟動「2030 永續發展目標」（Sustainable Development Goals, SDGs），提出 17 項全球政府與企業共同邁向永續發展的核心目標——第 17 項為「強化永續發展執行方法及活化永續發展全球夥伴關係」。這項目標的條例與核心精神是什麼？國內外又有哪些實例與反思？

17.1　強化本國的資源動員，作法包括提供國際支援給開發中國家，以改善他們的稅收與其他收益取得的能力。

　　17.1.1　政府總收入佔國內生產總值比例，依來源分列。

　　17.1.2　國內預算來自國內稅收之比例。

17.2　已開發國家全面落實他們的 ODA 承諾，包括在 ODA 中提供國民所得毛額（以下簡稱 GNI）的 0.7% 給開發中國家，其中 0.15-0.20% 應提供該給 LDCs。

　　17.2.1　官方發展援助淨額、總額和對開發度最低的國家的官方發展援助額，佔經合組織／發展援助委員會捐助國國民總收入的比例。

17.3　從多個來源動員其他財務支援給開發中國家。

　　17.3.1　外國直接投資（FDI）、官方發展援助和南南合作佔國內總預算的比例。

　　17.3.2　匯款總額（美元）佔國內生產總值的比例。

17.4　透過協調政策協助開發中國家取得長期負債清償能力，目標放在提高負債融資、負債的解除，以及負責的重整，並解決高負債貧窮國家（HIPC）的外部負債，以減少負債壓力。

　　17.4.1　債務還本付息額佔商品和服務出口額的比例。

17.5　為 LDCs 採用及實施投資促進方案。

　　17.5.1　國家通過和實施對開發度最低國家的投資促進制度。

17.6　在科學、科技與創新上，提高北半球與南半球、南半球與南半球，以及三角形區域性與國際合作，並使用公認的詞語提高知識交流，作法包括改善現有機制之間的協調，尤其是聯合國水平，以及透過合意的全球科技促進機制。

> 17.6.1　國家間科學和技術合作協議及方案的數量，依合作類型分列。

> 17.6.2　每 100 位居民固定的網路頻寬用戶，依速度分列。

17.7　使用有利的條款與條件，包括特許權與優惠條款，針對開發中國家促進環保科技的發展、轉移、流通及擴散。

> 17.7.1　為促進環保科技的發展、轉移、流通及擴散，向開發中國家提供之資金總額。

17.8　在西元 2017 年以前，為 LDCs 全面啟動科技銀行以及科學、科技與創新（STI）能力培養機制，並提高科技的使用度，尤其是 ICT。

> 17.8.1　使用網路的人口比例。

17.9　提高國際支援，以在開發中國家實施有效且鎖定目標的能力培養，以支援國家計畫，落實所有的永續發展目標，作法包括北半球國家與南半球國家、南半球國家與南半球國家，以及三角合作。

> 17.9.1　向開發中國家提供財政和技術援助，包括透過南北，南南和三方合作的金額（美元）。

17.10　在 WTO 的架構內，促進全球的、遵循規則的、開放的、沒有歧視的，以及公平的多邊貿易系統，作法包括在杜哈發展議程內簽署協定。

> 17.10.1　全球加權平均關稅。

17.11　大幅增加開發中國家的出口，尤其是在西元 2020 年以前，讓 LDCs 的全球出口佔比增加一倍。

> 17.11.1　開發中及 LDCs 佔全球出口的百分比。

17.12　對所有 LDCs，依照 WTO 的決定，如期實施持續性免關稅、沒有配額的市場進入管道，包括適用 LDCs 進口的原產地優惠規則必須是透明且簡單的，有助市場進入。

> 17.12.1　開發中、LDCs 及小島嶼發展中國家面臨之平均關稅。

17.13　提高全球總體經濟的穩定性，作法包括政策協調與政策連貫。

> 17.13.1　總體經濟資訊彙整。

17.14 提高政策的連貫性，以實現永續發展。

    17.14.1 國家制定永續發展政策協調機制。

17.15 尊敬每個國家的政策空間與領導，以建立及落實消除貧窮與永續發展的政策。

    17.15.1 發展合作提供者運用國家成果框架和規劃工具的程度。

17.16 透過多邊合作輔助並提高全球在永續發展上的合作，動員及分享知識、專業、科技與財務支援，以協助所有國家實現永續發展目標，尤其是開發中國家。

    17.16.1 國家為協助實現永續發展目標，在多方利害關係者發展的成效監測框架內，提出進展報告。

17.17 依據合作經驗與資源策略，鼓勵及促進有效的公民營以及公民社會的合作。

    17.17.1 投入公民營以及公民社會的合作之金額（以美元計）。

17.18 在西元 2020 年以前，提高對開發中國家的能力培養協助，包括 LDCs 與 SIDS，以大幅提高收入、性別、年齡、種族、人種、移民身分、身心障礙、地理位置，以及其他有關特色的高品質且可靠的資料數據的如期取得性。

    17.18.1 依據官方統計基本原則，由國家層級制定的永續發展指標比例。

    17.18.2 國家依循官方統計基本原則制定國家統計的立法。

    17.18.3 國家執行有充分資金的國家統計計畫，按資金來源分列。

17.19 在西元 2030 年以前，依據現有的方案評量跟 GDP 有關的永續發展的進展，並協助開發中國家的統計能力培養。

    17.19.1 為加強開發中國家的統計能力而提供的各種資源（以美元計）。

    17.19.2 (a) 過去十年進行了至少一次戶口及住宅普查，以及 (b) 出生登記達 100％ 及死亡登記達 80％ 的國家比例。

SDGs 的缺角　讓數位科技為人類和地球服務

　　2030 年是聯合國永續發展目標（17 SDGs）的驗收日程，隨著時間靠近，世界各國正在準備進行期中評估；同時，社會大眾也很好奇到底 COVID-19 疫情以後的世界，會變成什麼樣？儘管有許多不確定因素存在，但有一件事很肯定，我們的生活形式以及愈來愈多的經濟交易和政府事務將會轉移至網絡上。2015 年全球 193 個國家承諾致力於實現 17 SDGs 通過制定將人類健康、繁榮、環境、平權連接起來的變革性永續發展綱領。這個全面性的永續綱領，對全球民眾來說至關重要。2020 年初的新冠肺炎疫情危機，卻大大突顯出目前 17 SDGs 尚不完整。今日現有的 17 SDGs 尚未針對給予人類未來的嶄新定義的強大力量──「數位時代」（Digital Age），做出適當的回應與思考。

現況：數位落差加速社會風險與不確定性

　　數位時代正在重建舊有的社會結構，並且以前所未有的規模和速度，鋪天蓋地推動著變革。幾十年前，哲學家兼新聞學者馬歇爾‧麥克魯漢（Marshall McLuhan）提出格言：「人類創造工具，終而工具也創造了我們。」確實，現今的數位科技正在重新界定閱讀、消費、投票以及人們彼此之間的互動方式。許多風險和不確定性也隨之出現，包括對個人權利、社會平等和民主的威脅，而這些風險和不確定性都被「數位落差」（Digital Divide）更加倍放大了。數位落差是指網路普及的程度，以及接近使用數位資訊或產品之機會與能力上的差異，導致貧富差距更大。

我們倡議： SDG18——數位科技為人類和地球謀福祉

伴隨著巨大的危害風險，隨之而來的是偌大的機會。我們主張，可以善用數位時代更廣泛的影響力，來引導社會朝著共同的 SDGs 邁進，包括實現零碳排放和更平等的社會。在數位時代，透過科技我們可以將權力去中心化，並且轉移給其他需要的利害關係人；同時，也可以將社會消費習慣轉向低碳足跡產品；更可以將社會思維從使用化石燃料轉向使用再生能源。因此，我們需要一項針對數位時代的新永續發展目標——SDG18，以善加利用這一股強大的變革力量，造福人類和地球。

國際化一直是國與國之間的對話與經商必備的共同標準與共通語言，國家的經濟命脈學無止盡，臺灣企業 CSR 的績效已具世界級水準，但卻是呈現 M 型發展，高達八成以上中小企業仍未有 CSR 觀念或未執行及未落實 CSR 議題，薪傳智庫數位科技洞悉臺灣傳統觀念不擅於表達與故事行銷，殊不知每家企業在原有的行為與行動中已是 CSR 的體現，現在開始透過薪傳智庫數位科技 SESP 平台，簡易快速理解協助更多中小企業，運用數位科技讓中小企業的

小資源也能做大 CSR，亦是薪傳智庫落實 CSR 第 18 項指標：正確運用數位科技促進人類和平福祉的核心價值，透過數位知識與科技智慧的累積是人類重要的共同資產，並且可以確實持續不斷的更新。同時結合人工智慧及富有同理心、意向性的人類智慧、社會智慧，利用數位化潛力和相關的新知識財富，應對 21 世紀人類的重大挑戰，讓教育體系和全球知識轉移根本性的轉型。奧地利應用系統分析研究院、斯德哥爾摩韌性研究中心、聯合國永續發展研究網絡等三個知名永續發展領域的研究機構呼籲，全球應以聯合國《2030 永續發展議程》為基礎，即刻展開關鍵轉型行動。2019 年再以《數位革命與永續發展：機會與挑戰》為題，開啟關鍵轉型面向如何藉由數位工具加速落實。根據聯合國祕書長經過與各國政府、私營部門、民間社會、國際組織、學術機構、技術界等關鍵利益攸關方進一步展開一系列的圓桌討論，聯合國設想採取以下八大行動：

1. 在 2030 年實現全球連接。
2. 促進數位公共物品以創造更公平的世界。
3. 確保所有人，包括最弱勢群體，享有數位包容。
4. 加強數位能力建設。
5. 確保數位時代的人權保護。
6. 支持全球人工智慧合作。
7. 促進數位環境中的信任和安全。
8. 建立更高效的數位合作架構。

　　國際趨勢俾使薪傳智庫數位科技期待透過科技系統，可以共同將權力去中心化，並且轉移給其他需要的利害關係；這數位革命，包含了虛擬實境和擴增實境、3D 列印（或譯增材製造）、通用的人工智慧、深度學習、機器人學、大數據、物聯網和自動決策系統，在許多國家已進入公共辯論階段。回顧過往，幾乎難以相信《2030 永續發展議程》或《巴黎協定》卻鮮少提及數位化。顯而易見的是，「數位變遷」（Digital Changes）正成為社會轉型的關鍵驅動力。故，不少企業、政府、公民團體與學術界於 2020 年 6 月共同發表了一項聲明《數位時代永續發展蒙特婁宣言》（*The Montreal Statement on Sustainability in the Digital Age*），為倡議 SDG18 奠定了基礎。宣言希望社會明白到，應對氣候危機、建立全球永續發展、實現平等的數位未來，是該努力並且是互相關聯的議題。邁向永續的轉型必須調和其威脅、機會、數位革命的動力，同時，數位轉型會徹底改變全球社會與經濟的所有面向，因而改變永續性本身的典範詮釋。

數位化是一個跨尺度的經濟、社會和文化連結，結合真實和虛擬情況的強大乘數，更重要的是，創造出強化人類感知，具認知能力的技術系統（例如人工智慧和深度學習）的特性，至少在某些功能領域裡，最終將補足，或許甚至會取代或最後遠勝人類的認知能力。數位化不再只是解決永續性挑戰的一種「工具」，它同時也是具破壞性、多尺度變化的重要驅動力，為了永續未來的重要問題，也需要因應數位化，使其帶來最大的機會與最小的風險，國際上提出連結數位動力和永續策略的六大重要機制：

1. 在科學、研究和研發社群中創造永續數位化的視角，改變創新的願景和模式。

2. 制定適當價格來驅動市場力量，例如透過碳定價和生態稅改革，來激勵數位創新行動，以支持永續的解決方案。

3. 運用數位化來具體化和制定轉型路線圖，包括更明確地界定能源、交通、土地利用系統、城市和工業部門的目標和里程碑，幫助市場及規劃進程轉往更永續的方向。

4. 從國家層級投資數位現代化計畫，大規模增加公立機構中的數位知識，藉以建立數位人類世的治理能力。

5. 透過支持和擴大與數位研究社群的強大網絡，建立具有變革性的永續研究。

6. 建立與私部門、公民社會、科學和國家的對話機制，共同探討發展數位人類世的制度、社會和規範保障。

數位化將是新的經濟、社會與文化的現實和挑戰，虛擬實境、人工智慧、深度學習、大數據以及愈來愈多用於規劃與營造情境過程的活動，改善我們在複雜的社會生態系統中理解決策意涵的認知能力，知識突破和爆炸會為人類帶來空前的新潛力。決策者、研究者、企業和民間社會行動者必須加強努力，去理解和解釋數位變遷的多重影響、預料深遠的結構性變化，善用科技做為觸發各領域投入的火種，可以擴大科技政策的外溢效果，使其創造出塑造數位化進程的基礎，合適的數位化進程和相關科技（如新複合材料、奈米技術、奈米生物技術、基因工程、合成生物學、仿生學、量子計算、3D 列印以及人類增強技術（Human-enhancement）增進技術（Augmentation）將使智人轉變為數位人（Homo Digitalis），人工智慧、深度學習和大數據將改變科學，為人類文明的下個階段敞開新大門，虛擬存取有關人類和地球最先進的全球知識，可以為所有人實現公平、公正且安全的未來。以及透過數位動力能促進文化、制度和行為創新，跨國的傳播網絡有助於建立全球網絡型社會、跨國治理機制、全球共有財觀點、全球合作的文化和跨國認同，也可能創造新的（次）文化，虛擬網路中來自世界各地的人們，可能會改善我們對文化多樣性的理解。虛擬實境能讓人類不需長途跋涉，即可「造訪」、瞭解、享受以及「感受」全球的生態系統。同時，透過數位投票程序，促進民主發展機制的新選擇也正迅速擴展，包括線上查核當地決策是否涉及與轉型和改革偏好有關、實際但重要的治理問題。綜觀以上，在數位技術和匯流必定增強及增進人類的機能和認知能力，驗證上個世紀的人類機能已有大幅改善，醫療健康、競技運動和知識方面都取得了前所未有的成就，讓人的壽命在上個世紀間已成倍增長，透過數位的增強和增進或許會更無限延長，人造器官和義肢的使用也將開始經歷重大突破，像是外骨骼強化和人體機能增進的全新改良，以及機器製造機器的古老科幻願景已經成真，使人擺脫身體勞動，增進及增強認知與體能。這些新興創見表現出人類新時代的潛在正向性，數位化提供了極佳的可能性，並使之朝永續轉型發展。

（三）2021 是 ESG 發展元年帶來轉機與希望

2021 是 ESG 發展元年，國家、產業正視其重要性帶來轉機與希望，啟動改變與新機遇成為首要之務，「2021 遠見 ESG 永續金融論壇」即深度探討2021 年追求永續議題，金融發展的未來關鍵。《遠見雜誌》社長楊瑪利指出，在金管會政策的推動下，ESG 議題在 2020 年謂為顯學，2021 年更將會是 ESG發展的發展元年，各產業無不積極投入此一議題、蓬勃發展；而當前的數位化挑戰是要解決數位人類世下艱鉅的永續性問題，扭轉形勢成為當務之急。在2030 永續發展議程視角下的急迫需求，距離 2030 僅剩不到 10 年，社會及其政府正處在最重要的關鍵時期，我們明白形塑和治理邁向永續的數位革命沒有萬靈丹，因為未來本就是不確定的，時間是非常寶貴且稀缺的資源，我們必須明智地運用。我們不到 10 年時間能全力動員，並借助數位機會創建出永續社會，並且去習得如何管理數位的綠色經濟與穩定、公平、開放的數位社會，正面運用數位化與人工智慧的社會影響力，融合虛擬與實際的空間和真實情況，避免進一步削弱社會凝聚力。未來的挑戰無疑會是認知上的強化，這個挑戰在於建立有韌性、適應性、知識性和具包容性的「責任社會」，數位化在教育、衛生、公平和繁榮方面取得重大進展的潛力是不容否認的，對於生活與工作的方式及地點、如何利用增加的休閒時間，以及如何與即時、在地和更廣泛社群的其他成員互動，同時理解人工智慧、自動決策系統和虛擬空間等新興挑戰，這些方式的重大變遷將帶來社會影響。薪傳智庫數位科技平台希望匯集數位及永續研究社群的觀點，共同與國際夥伴攜手利用數位化、虛擬實境和人工智慧的機會，遏止其潛在風險並連結數位和永續轉型。跟隨以下基石創造了相互依賴的系統架構，以助於管理數位和永續轉型的調和化：

1. **教育：**人們需要能理解並形塑新興的「數位轉移」（Digital Shifts）。
2. **科學：**新知識網路必須創造轉化型知識（Transformative Knowledge），來整合數位和永續導向的轉型、避免逼近數位臨界點，並建立整合人類和智慧機器時代的規範性框架。
3. **促使國家現代化：**公立機構尚未準備好理解與治理數位動力。大規模現代化和教育計畫在這方面是必要的。
4. **實驗場域：**邊做邊學和應用，尤其在早期的創新階段，是科技和制度傳播的主要原則。應該建造些創意場域來培養快速學習，並包含能讓「狂點子（Crazy Ideas）和新創企業」落地的可能性。
5. **全球治理：**數位革命在建立聯盟方面有全球影響力。舉例來說，聯合國的現代化將受到數位化時代的形塑。
6. **「新人文主義」**（WBGU, 2019）：《2030 永續發展議程》可以被視為對世界的新「社會契約」，它改變了我們對 2030 年以後的未來價值觀與願景，並朝著整體永續發展。這意味著人類和地球有對未來的新規範目標，唯物主義、環境與地球系統負面外部性脫鉤的新發展模式，及遍及眾人的新規範性保障。

　　COVID-19 疫情讓世界正處在十字路口，SDGs 應該被視為 2050 年及其日後實現永續發展的關鍵重點，臺灣以中小企業為主，在面對轉型升級、數位人才、軟體資源不足、上下游資訊缺乏、國際市場開拓問題、大數據、AI、5G、IOT 等多種新科技尚未整合下；讓邁向永續發展的轉變需要很長時間才能完成，現在亟需有相對應的管理政策、獎勵計畫以及價值觀轉變，然而事實顯示全球只有僅存在於少數幾個國家，為了實現這六大關鍵轉變，推動連結數位動力和永續策略的六大重要機制。

　　這場全球疫情，不但燒出了世人對於高度倚賴經濟所衍生出來的諸多地球問題，對企業而言，疫情是危機，也是轉機，更對永續發展帶來常態性的改變，永續投資成為不可逆轉的大趨勢，CSR 方法已從臨時、增量和交易方法轉變為戰略、社會目的驅動和轉型模式。CSR 正在經歷從「精打細算」的驅動力向「企業成功必不可少」的動機轉變。薪傳智庫數位科技平台讓每個特定的夥伴關係都是獨特的，需要合作夥伴來促進創新和取得成功，他們的共同點是：CSR 承諾符合公司的基本價值觀和策略方向，因此是值得信賴且真實的。

　　根據麥肯錫研究，到 2025 年數位經濟對全球 GDP 的貢獻約佔 49%。而臺灣 2025 年數位經濟 GDP 佔比目標為 29.9%，參考主要經濟體數位經濟發展速度，推估臺灣 2030 年數位經濟 GDP 佔比將接近四成。雖然永續投資已經談了十幾年，真正風起雲湧卻是近期的事。整個投資產業價值鏈中，利害關係人包括：客戶、銷售、合作夥伴、員工、股東、所投資公司、供應商、監管機構、評級機構以及媒體、當地社區、非政府組織等民間團體，此一趨勢帶動下的企業價值無形中也正在被重塑，永續投資成了幫助企業提升價值的關鍵觸媒。一場擴及全球、左右未來競爭力的企業淘汰賽，已悄然登場，門票正是 ESG，企業沒有 ESG，恐怕連登上擂台，與同業競爭的資格都沒有。有研究指出企業對於財報會議中不僅對疫情有大量的討論，對於非財務報告更是大幅增加，ESG 與 EPS 一樣重要，必須由上到下、由內到外齊心協力貫徹執行，企業亦重新定位營運模式、檢視風險因子和控管；對於地球到這片土地，乃至於與員工之間的多元化與包容性、健康與安全、薪酬福利、勞工關係及平權等面向，故，薪傳智庫數位科技平台擴散分享政府美意，協助推動勞動部就業相關補助（青年、職場再適應、雇用獎助、中高齡、安心就業、職務再設計、產學合作）、活化與提高就業人口、人力提升計畫補助推動員工職能發展與工作效益、工作生活平衡補助，提升重視員工的工作環境與福利，大大降低員工流動率，讓工作效率大幅提升，除了促進企業與員工誠信、忠誠度等工作態度，溫馨更烙印企業形象建立和凝聚員工向心力，以及協助企業建構人才培訓體制 TTQS 企業版與機構版並取得國家牌等，甚至依企業人才特性、級別規劃 iCAP 職能導向課程，皆依企業核心發展轉化為 CSR，趨動企業去應變市場的挑戰，從長遠來看也將更顯而易見。COVID-19 疫情因素影響下，人類更深刻體會 ESG 是永續未來的重要關鍵，從個人到企業、國家都該挺身而出，以實際行動投入，才能扭轉危機，成為能讓世界共融共好的轉機。

以 G20 高峰會預估資料為例，若要在 2030 年前完成聯合國永續發展目標，必須在永續能源基礎上再投入 2.9 兆美元；放眼整個資本市場已將 ESG 納入投資的標配流程，依據投行 BofA 預測，未來 20 年 ESG 基金管理的資產將達 20 兆美元以上，瑞銀更預估 2036 年 ESG 基金規模上看 150 兆美元，重點在 ESG 投資績效：法巴能源轉型股票基金，半年內猛衝 174%，還有王道銀行推出全臺首見的「影響力存款」專案，CSR 讓理財也能積福報。再加上 2020 年 12 月 ESG 揭露：臺灣 ESG 是居亞洲之冠，臺灣政府帶頭在 2021 年 3 月也正式計算「辦理綠色採購」分數。所以，企業要能在 ESG 上受到資本市場金融青睞與獲得更多合作及採購機會，就勢必將 CSR ／ ESGs 內化為公司的創新與成長策略，但產業與文化不同的 CSR 行動讓中小企業執行內容與項目相當混亂，這再次證明社會需要企業的解方，薪傳智庫數位科技平台驅使與誘發企業發揮更大能量，將「個制化」提供企業創新營運模式應用實務找到出路的曙光機會，落實經營方針規劃 CSR 的具體策略，參與透視各利害關係人從理解到落實 CSR 過程中事實證明，透過深度溝通企業語言與善盡落實 CSR 計畫執行，取得政府各部會之補助與科專計畫及紓困陪伴指導下，讓企業跟進政府政策下共渡難關，手心向下重視 CSR，以行動執行力倡導國家政策，協助企業疫情紓困補助與紓困貸款、青年創業和微型鳳凰以及國家投資「臺灣三大投資方案」投資臺灣事務所各項溝通協助；提高與同業競爭的門檻；讓一個支點帶動加乘的國際經濟效益，積極打造創新的價值鏈產業環境。如同 Apple 宣布 2030 年要達到百分之百碳中和，臺灣很多廠商都是 Apple 的供應鏈，面對接下來 10 年綠色供應鏈將有爆發性成長，生活中的用品，從衣服、眼鏡到鞋子所使用的材料會發生巨大的變化，整個產業的供應鏈將轉型升級至循環永續的體系；臺灣在循環經濟、永續創新方面具有極大優勢與商機。薪傳智庫數位科技願臺灣的國際競爭力，以及臺灣的對外關係，形成邁向一個新世代的新國力。

資料來源

- 天下雜誌（未來城市編輯部），SDGs 目標 1～17，取自 https://topic.cw.com.tw/event/2020sdgs/
- 未來城市編輯部（2021 年 1 月 27 日）。SDGs 懶人包》什麼是聯合國永續發展目標 SDGs ？一次掌握 17 項核心目標，取自 https://futurecity.cw.com.tw/article/1867
- 朱竹元（2019 年 5 月 16 日）。【專欄】朱竹元：永續經營　不該是有門檻的事，取自 https://csr.cw.com.tw/article/40963
- 李昕蒨（編譯）（2020 年 10 月 14 日）。蒙特婁數位時代永續聲明，取自 https://rsprc.ntu.edu.tw/zh-tw/m01-3/understand-risk-society/295-sustainable-tra/1488-1014-systemdynamic.html
- 林千慈（翻譯）；郭映庭（審校）（2019 年 11 月 13 日）。TWI2050 報告（2019）《數位革命與永續發展：機會與挑戰》重點摘譯，取自 https://rsprc.ntu.edu.tw/zh-tw/m01-3/tech-pros/1319-1081113-key-factor.html
- 美國在臺協會，聯合國永續發展目標（SDGs）說明，取自 https://www.ait.org.tw/wp-content/uploads/sites/269/un-sdg.pdf
- 風傳媒（2020 年 8 月 27 日）。為何全球資本市場日益重視 ESG ？投資專家：數字會說話，取自 https://www.storm.mg/article/2980627
- 陳清祥（2014 年 7 月）。如何全面性實踐企業社會責任，取自 https://www2.deloitte.com/tw/tc/pages/risk/articles/newsletter-07-6.html
- 黃曉雯（2021 年 3 月）。從財報、股東、資本市場 看永續治理新貌，取自 https://www.accounting.org.tw/flptopic.aspx?f=306
- 顧燕菁（2021 年 1 月 29 日）。攜手投入永續發展，是扭轉可能風險的集體共識，取自 https://www.gvm.com.tw/article/77561
- Amy Luers（作者）；CSRone 鄭鈺弘（編譯）（2020 年 11 月 22 日）。【CSRone】SDGs 的缺角：數位科技為人類和地球服務 SDG18。取自 https://csr.cw.com.tw/article/41743
- Apple（2020 年 7 月 21 日）。Apple 承諾要在 2030 年對供應鏈和產品實現 100% 碳中和。取自 https://www.apple.com/tw/newsroom/2020/07/apple-commits-to-be-100-percent-carbon-neutral-for-its-supply-chain-and-products-by-2030/
- NTU SDGS（2020 年 1 月 2 日）。想像 2050- 落實永續發展目標的關鍵轉型行動。取自 https://oir.ntu.edu.tw/ntusdg/outcome/%E6%83%B3%E5%83%8F2050-%E8%90%BD%E5%AF%A6%E6%B0%B8%E7%BA%8C%E7%99%BC%E5%B1%95%E7%9B%AE%E6%A8%99%E7%9A%84%E9%97%9C%E9%8D%B5%E8%BD%89%E5%9E%8B%E8%A1%8C%E5%8B%95/
- The United Nations Statistics Division. SDG Indicators. 取自 https://unstats.un.org/sdgs/metadata/

四、CSR 實務行動

（一）教育推動篇

大學社會責任對臺灣企業永續發展的重要性

國立高餐大系主任
梁榮達教授

當 CSR 的實踐範疇從企業擴大到社會各角落，大學也開始推動大學社會責任（USR），透過大學的專長與優勢，引領年輕學子一同為社會帶來改變。自 2016 年來，教育部積極推動 USR 的執行，引導各大學跳出傳統的教學框架、提升與地方社會的互動及連結，並培育出符合社會發展需求的人才；從而發展出各個學校的特色、瞭解社會責任的意識，亦解決存在於社會各個角落的問題。這些問題的根源，可能是行銷上的困難、專業能力的不足，或者是特色的缺乏，而這些都是大專院校自身具備的能力，專業的學術師資，富有創意與行動力的年輕學子，此時若結合 CSR 的資源，以 CSR×USR 的加乘概念，從企業投入及學生教育等兩大面向著手，全力推動下的成果與效益將更為可觀。

許多企業主認為推動 CSR 只是花錢，像是捐錢的慈善活動，其實不然。CSR 的面向極為多元廣泛，例如提供員工完善的工作環境，協助鄰里開發社區在地特色。更重要的是，發揮企業專長，以特色資源來營造難以模仿、容易永續發展的優勢。CSR 架構分成三個階段：第一，開始階段，是內外部力量趨動 CSR 行動階段，開始的動機來自於公司社會規範、政府政策、領導價值及組織文化。第二，執行階段，企業開始運用：(1) 經濟策略，例如慈善事業、永續觀光產品；(2) 環境策略，例如節能減碳、環境保護、永續資源管理；(3) 社會策略，例如顧客教育、產學合作；(4) 文化策略，永續文化遺產或支援文化及創意產業。第三，回饋階段，企業可能得到某些良善成果，例如正面回饋，企業形象提升、內部及外部顧客的滿意或是追求永續經營。當中，社會策略中的──產學合作，便是此篇章欲闡述的主軸，我將以過去曾擔任教育部大學社會責任計畫主持人及協同主持人的經驗，輔以我在高雄餐旅大學實務推動的產業研究成果分享，呈現 USR 如何積極輔導社區協會營造出社區特色，以學界的角色輔導地方社區及企業共同發展，創造雙贏，攜手達成永續目標。

　　高雄餐旅大學一直以來是大家熟知培育餐旅人才的搖籃，更是餐旅企業的關鍵夥伴，學校長期致力於推動師生實踐、創新創業、地方創生與產業國際化。在與高雄市橋頭區區長陳振坤的合作下，我們串連起高師大、高科大（第一校區）、朝陽科大的夥伴學校，獲得教育部 106 年度「大學社會責任」（USR）實踐計畫「"傳" 到橋頭自然 "職"」計畫補助。本計畫核心價值為：運用高餐大餐旅創新專業、夥伴學校的經濟、行銷及農業專長，結合在地社區及產業力量，推動橋頭在地的濃（酪農）、醇（蜂蜜）及香（瓜果）產業特色。計畫成員認為，透過產、官、學界的資源及力量，建立出具有在地特色的景觀及食材，協助地方提出具有當地食材特色之農特產商品、在地景觀之遊憩行程，可以解決橋頭區區域經濟發展問題。透過農特產品開發／有機輔導／青農行銷；其次為開發行銷曝光增加橋頭特色媒體能見度。長期目標為營造專屬橋頭甜蜜社區的特色人文風貌，其中包含了讓在地食材轉型成在地伴手禮、在地農業轉型為在地休閒農業，以利區域轉型及特色營造，進而讓橋頭區成為高雄的「甜蜜後花園」。由於在地人民因為產業多以生產導向的經營方式，忽略了提升附加價值的機會。因此，團隊成員協同學生規劃特色伴手禮以及社區遊程，盼望能解決核心問題。團隊妥善結合了橋頭區的牛奶、蜂蜜及火龍果，開發出火龍果系列的果醬、雪 Q 餅、果汁氣泡水、冰淇淋與霜淇淋，做為橋頭水流庄社區的甜蜜特色食品。而本計畫在執行後：第一，針對地方資源不足的情況，透過成員及學生的努力，協助地方政府加強經濟發展，以及地方特色的塑造。第二，在多次接觸與師生參訪，形成了與社區合作互動的默契，並且由計畫提供實質的創意發想、協助社區體驗活動的開發，社區環境維護等，解決了人力及物力不足的問題。第三，對於地方產業而言，透過利用在地食材開發出的特色商品，有助於解決食材供給過量，同時營造特色文化的利基。而這樣的 USR 計畫不僅是大學應用專長來幫助社區，更是將在社區所做、所學，應用在大學教學現場，解決現今學生學習動機不足等問題。

　　個人相信如同政府在推動大學落實社會責任實踐的理念，透過強化大學與區域城鄉發展的連結合作，讓大學走出學術象牙塔，在區域創新發展的過程中，對在地產業、社區文化、城市與偏鄉等發展議題，投入更多學界能量深耕在地，讓大學成為地方永續發展的積極貢獻者，協力臺灣社會共同成長。而政院推動 USR 至今，正邁向「地方創生」與「國際連結」這兩類議題，以振興、創新產業，吸引人口回流為目標，串連世界、穩步投入國際參與，也正是這個擁有無限可能的時代，驅動 USR 持續為社會永續發展帶來改變的動能。在 CSR 的影響力逐漸擴大到眾所矚目的時刻，USR 的崛起成為追求永續發展目標路上的續航力，我個人過往的實務經驗與成果分享，相信可以做為更多社會進步及學生實務學習良好的教材！

（二）空氣循環篇

氣候變遷與「氣候行動」環境永續

威浦斯科技股份有限公司董事長
王雪慧

空氣是一個人生存不可缺少的元素，但根據 WHO 調查統計指出，目前全球約 700 萬人的死因為室內外的空氣污染。無論市區還是鄉村，空污都已成為環境風險因素的榜首。雖然目前大眾在公共衛生的危機意識已大大提升並努力落實，但是我們必須更深入瞭解空污對地球與人類所造成的深遠影響。據研究指出，空氣污染所損害的領域早已超越個人健康，並延伸至氣候變遷、乾淨水源、再生能源和農業發展議題。

「陽光、空氣、水」是人類生存最基本的三大要素。然而，現今受到污染的空氣數量卻已多到能夠輕易地損及人類「從頭到腳」的每一個器官與細胞。以一個成年人做為衡量基準，平均每分鐘會吸進、吐出 7 到 8 公升的空氣，等同於每天至少呼吸 11,000 公升的空氣，無論它是否乾淨。我們都知道空氣污染會損及肺部與氣管功能、加速生命邁向死亡。不過，據國際呼吸學會論壇（Forum of International Respiratory Societies）研究人員於《胸腔》（Chest）期刊指出，污染加劇的空氣吸入人體後所影響的範圍將會更廣泛。除了心肺血管

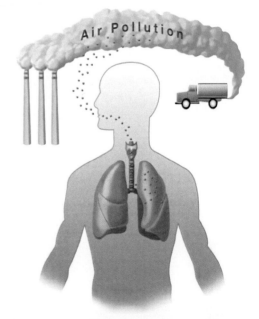

疾病之外，還可能導致糖尿病、失智、膀胱癌、骨質疏鬆症、生育能力降低等疾病不勝枚舉，尤其是肺癌在臺灣癌症排行榜中，已連續 40 年都是在前兩名之中，可見空氣污染早已擴散到影響人體健康，而不自知。然而，在 2020 年引發的新冠病毒，美國疾病管制暨預防中心（CDC）在此次的 COVID-19 中表

示，病毒能藉由懸浮於空氣中可達數小時的病毒傳染，承認公衛專家普遍提出有關病毒恐透過空氣傳播的隱憂確實存在。

威浦斯科技在能源與環境資訊發展中心翁教授的努力之下，在這疫情爆發的防疫時代，「全外氣自然能轉換系統」扮演了重要角色，全球首創氣流分隔能量轉換技術，採用抗鏽防腐隔板材質，使進氣與排氣的流道完全分隔，在進行能量轉換時，進行滅菌消毒功能，徹底解決空氣交叉污染之問題，杜絕疫情擴大的機率，適用於醫療產業，另可視需求，設定為「負壓環境」或「正壓環境」，已應用於彰化秀傳紀念醫院與桃園敏盛醫院，可謂「最佳防疫工具」。且該套系統透過建築物內外水蒸氣分子壓力差進行能量轉換，一來可有效降低二氧化碳濃度，二來亦可減緩地球溫室效應的速度，除了避免空氣交叉污染，擴大疫情，在能量轉換中，不僅不會排放熱氣，還可降低二氧化碳濃度，提供產業建築物以最低能耗，創造類自然室內環境更是緩衝了地球的溫室效應速度。更是在 CSR 18 項指標中，將此技術奠基於臺灣，而發揚光大於全球，造福更多的生命與環境，努力落實 SDGs 中第 13 項指標的氣候行動，共同與薪傳智庫數位科技股份有限公司為 2050 年的目標付出一點心力。

氣流分隔材質：抗鏽防腐隔板

利用機體結構的抗鏽防腐分隔板，使進氣與排氣的流道完全分隔，在進行能量轉換時，無空氣交叉污染之問題。

（三）海洋公約篇

綠色和平全球海洋公約

漁業
林天印董事長

　　海洋是生命之母，蘊孕生命的搖籃，無論生物的演化論，或者是航海大發現，海洋對人類的歷史發展重要性是不可言喻。海洋對人類的價值除了調節全球的氣候，創造人類賴以生存的自然環境；海洋擁有豐富的生物資源，亦是人類的重要食物來源。海洋是生命的誕生和孕育之地，它不但佔了地球表面71% 面積，生物棲地體積的 99%，在地球循環及供應，生命也才得以在這個世界生生不息，這個重要的循環過程，同時更在人類文明的演進中扮演著重要的角色。

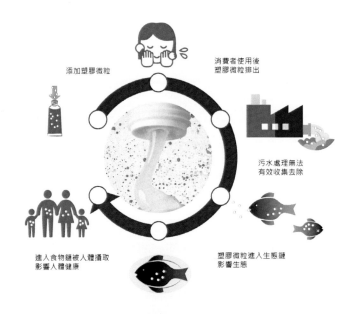

　　臺灣自命為海洋國家，但政府對海洋做得太少，水域生態系統兩種主要類型是海洋生態系和淡水生態系統，世界知名的魚類中有41% 棲息在淡水生態系統中。而臺灣人對大海太過無知，對多數人而言卻無比陌生的大海，史

上首次跨季度海洋調查：橫跨四個季節的初探調查結果總算出爐，大海，在求救。訴說著大海如何被塑膠垃圾攻占的悲劇。美國海洋學者 Jenna Jambeck 統計，每年大約有 800 萬公噸的塑膠垃圾被排入海洋，為海龜、鯨豚、鳥類等將近 700 種生物帶來生存危機。隨著海洋污染物越來越多，由塑膠製品裂

資料來源：The Ocean Cleanup。

解後所成為的塑膠微粒，這些碎片不僅容易吸附環境中的重金屬、戴奧辛，也更容易被生物誤食，已經進入食物鏈，成為我們餐桌上的佳餚。台灣清淨海洋行動聯盟（TOCA）曾與國際學者合作，分析臺灣民眾 12 年來 541 次淨灘活動的成果，發現臺灣海灘上的垃圾，有 90.8% 為完全塑膠或部分塑膠材質，最常見的前 5 項垃圾分別為：1. 塑膠提袋；2. 塑膠瓶蓋；3. 免洗餐具；4. 漁業用具；5. 吸管。黑潮採集塑膠垃圾的方式，是採用美國 NGO 組織 5 Gyres，再進一步說明採集來的塑膠垃圾分為以下 5 類：1. 硬塑膠；2. 軟塑膠；3. 發泡塑膠；4. 塑膠纖維；5. 圓形塑膠粒。

英國研究全球每年有 1,270 萬公噸垃圾進入海洋，其中 94% 沉入海底，剩下 6% 才是沙灘，世界經濟論壇警告，如果我們對於塑膠垃圾的控管沒有改變，到 2050 年時，海洋中的塑膠垃圾總量將比魚還多。環境顧問業者結合國內海洋學者公布臺灣海底垃圾調查，發現西海岸八大外海的海底垃圾密度是全球的 1.5 倍，逾九成是纖維布料及膜狀塑膠。海底垃圾過去甚少被關注，外界多關注沙

灘、海面等肉眼可見海廢，學者示警，海底垃圾不僅影響生態，進入魚類、浮游生物身體後，最後恐透過食物鏈被人類吃下肚。

11 個寶特瓶可製作一雙全新運動鞋

7-8 個寶特瓶可製作一件全新T-shirt

1-1.5個寶特瓶可製作一支全新寶瓶

　　以撿垃圾又做公益的活動去幫助更多人，為一舉兩得之見，海洋中塑膠垃圾現存量超過 1.5 億噸，如果我們不採取行動，10 年後的 2030 年，海洋垃圾將是現前的 2 倍。直到 2025 年，每 3 噸魚將含有 1 噸塑膠，2050 年，海洋中的塑料將比魚多（按重量計）。全世界都重視海洋污染，桃園海岸線 46 公里，市府截至 2020 年 9 月底，辦理淨灘 71 場次，超過 1.3 萬人次清理海岸垃圾，總共清掉 218 噸垃圾，桃園市府也與在地廠商合作，使用回收寶特瓶製作環保鞋子、運動 T 恤產品，2020 年更研議收購漁民廢棄漁網、啟動漂流木再利用示範計畫，收集漂流木製作再生傢俱或木棧道鋪面，希望海洋垃圾回收再利用，減少海岸垃圾。和泰 TOYOTA 集團 2020 年 10 月 18 日上午舉辦淨灘減塑全臺總動員，發動 1 萬多人前往桃園市觀音區濱海遊戲區沿岸撿拾垃圾、廢棄寶特瓶淨灘，市長鄭文燦表示，和泰汽車再次在全臺 17 處展開淨灘，每撿拾 1 支廢棄寶特瓶捐 10 元，所得總額全數捐贈台灣咾咕嶼協會，做為全臺國小孩童減塑環保教育，落實 CSR，成為環保企業標竿。《KPMG 全球企業永續報告大調查》結果顯示，全球（63%）及亞太區（66%）超過六成的企業 CEO 表示，在全球環境充滿變數的情況下，將更加重視 CSR 及 ESG 議題，而臺灣也有過半（52%）的 CEO 認為疫情更加突顯了 ESG 議題的重要性，企業如何在新的一年，突破困境、開創企業永續新局，是當前許多企業面臨的重大課題。

　　為此我想代表漁業祈願我們一起為臺灣的海洋生態做些改善，期待從南部開始與薪傳智庫數位科技股份有限公司共同行動去達成社會企業責任（CSR）。經商往來日本的我提出日本小豆島與大家分享實務，默默無名的日本小豆島擁有滿滿大海的恩惠，是日本近世前的重要海運動脈，邁入近代後受到工業化、全球化衝擊，遂使瀨戶內地區的小島人口外流、島內高齡化加劇，是一個沒有任何資源的小島，讓這片海洋逐漸喪失了生機。**「海之復權」**是瀨戶內國際藝術祭的宗旨，在建築師安藤忠雄和藝術家草間彌生為了活化地方，開始計畫性復興瀨戶內海上的小島，每 3 年一次與海洋島嶼的約定，天使之路是一條沙路，連接四個小島，一個由海洋、島嶼、砂州組成的夢幻美景，讓既有自己的傳統，也有西方引進的技術，還有世界各地來的藝術發想，是傳統的小島、是走向世界的小島「希望之海」。

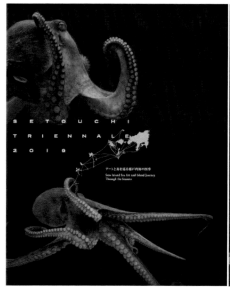

　　「海中生物」為主題，希望能呈現出在平穩寧靜的瀨戶內海之下，是一片未知的幽暗的海底世界，聚集著我們不熟悉的海中生物。原研哉以跳脫既定印象的手法，希望讓觀者心中浮現「？」、「！」的驚愕與驚嘆，並傳遞人海的深邃魅力。

（四）前瞻建設篇

前瞻基礎建設循環經濟

建設公司董事長
林國文

　　臺灣是個以外銷為主的國家，而土地是財富之母，但在國內的土地小再加上寶島土地資源有限，相較國際土地都是國有財，臺灣土地持有之「有土斯有財」的觀念及預期未來土地上漲的心理下，造就了建商投機與投資客之高獲利之工具。然而臺灣長期沿襲日本建築法規，在營建過程的建築結構仍無法突破，僅在建築設計採用歐美國家規範，在現今 RC 建築結構如何邁進國際的 3D 列印房，造福更多人民居住問題，同時廣泛的解決營建業缺工與人才及品質各項成本；不應羊毛出在狗身上豬買單，歐美國家與中國已在市場機制施行 3D 列印房多年，且運用更多資源回收再造（鋼渣、廢料、鐵泥……），此時，CSR 居住正義正是內政部營建署擺脫官僚最好的挑戰與最佳時機普及化，若透過建築結構不斷測試，將智慧城市技術創新創造體驗應用，再造臺灣營建業經濟動能。

資料來源：蕭羽耘（2020 年 12 月 03 日）行政院一張圖教你打房！　推「5 大面向」抑制炒房。取自 https://www.ettoday.net/news/20201203/1868549.htm

1. 建設公司一直是蘊育臺灣寶地黃金年代的火車頭產業，除帶動其他產業的關聯消費外，建築業滿足城市發展與商業發展環境的指標，基於人的關係和最高價值而存在，對於個體而言，社會是個溫馨的字眼，只是營建行業的存在與否以及發展方向，是依據市場的需求而存在與發展，然而卻在良莠不齊的建築業中，帶來家的幸福背後，背負著負面的社會觀感，在建築業生命週期過程中（包含規劃、生產、施工、使用管理及拆除、買賣與租賃等業務），產業價值鏈活動；以勞動密集型為特徵，銷售與獲利的成敗，具備相當程度的專業性，且必須擁有基本的技術團隊，產業特性易受政府政策及推動重大公共工程計畫之影響，也由於營建工程公司多為中小企業，營建產業被廣泛認為是道德聲譽低落並對生態環境造成破壞。臺灣在 2003 年 SARS 過後開始起漲，並展開了一段長達 12 年的大多頭行情，在 2014 年創下 14 年新高後，價量齊揚，加上投機客的炒房演變成土地炒作的助燃劑，間接更變成房地產無法帶動國內內需動能，反而變成內耗的產業，整個產業逼出了臺灣房市黑天鵝政策，接著政府祭出「房地合一稅」打房。

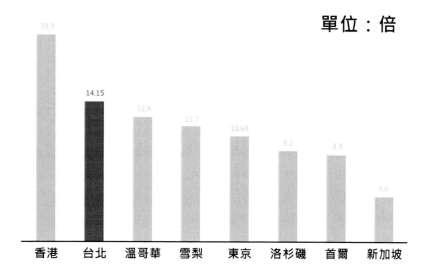

單位：倍

資料來源：美國國際公共政策顧問機構 Demographia 一年一度的《2019 年全球住房可負擔性調查報告》，但不包括中國、臺灣、日本、韓國等國家調查，因此臺北、東京、首爾採用當地政府資料統計。臺北為 2019 年第一季數據。

2. COVID-19 造成全球化影響力不只具現化於經濟市場中，臺灣的營建業更受到國際組織參與的限制導致落後國際，深思企業是社會的細胞，社會是企業利益的源泉，疫情中恐慌的資本投資潮流向 CSR，CSR 被視為疫情中最具有競爭優勢的潛在來源，是混亂、競爭和多變、不確定的環境中，一種最有效且必要的「經營戰略」。

   (1) 全球營建業推動落實永續責任之現況與趨勢，相比臺灣國內營建業在 CSR 領域窒礙難行。

   (2) 營建產業之 CSR 亦僅在起步階段，政府和企業對 CSR 的認知與積極度不足，成為臺灣企業在全球競爭的隱憂。

   (3) 雖然營建業有別於一般企業，特別是營建環境錯綜複雜，受到法令、環境、國情與人情、龐大資金等因素環環相扣，除要達到設計施工品質、安全要求，更需強化技術及創新管理能力以免被大環境淘汰，若能善盡 CSR 不只是政府與消費者的期待亦是營建業的期盼。

   (4) 遑論目前營建業景氣低迷，文化與脈絡下不同的 CSR 行為，現在消費者重視綠建築結合本業，目前營建市場有 APP 有智慧建築與無障礙環境建築的設計，有減少環境的負擔與能源的浪費，降低環境的成本；也有 ISO 管理系統及 CRM 客戶服務系統，以及家的健康品質證明書，還有 ICT 物聯網建構客戶全方位的精緻價值鏈、綠色供應鏈聯盟，木屋建材市場、SRC 鋼構建築雨後春筍般相繼投入藝術……等等；都是需要長期推動，並非一時急就章可達成，但有助於調整消極與被動之心態。

3. 2050 年，全球的基礎設施將成長四倍，展望未來趨勢「下一波創新的浪潮」我認為 3D 列印房子，才能真正解決房價與居住正義。以國外歐美為例，甚至在中國，營建業產業實現 3D 列印房子與列印公共建設發展的樞紐，不僅快速解決缺工與技術品質問題，若我們從營建業 3D 列印房子與公共建設是國家發展的引擎，短期思考如何技術創新，不但能在將來市場與國際對接具有刺激景氣、增加就業與內需的功能，同時以整合及旗艦的策略，銜接臺灣的優勢與機會，創造生活產業的體驗應用，長期更可蓄積資本存量，促進民間投資及改善國民生活品質，強化亞洲和全球的競爭優勢再造臺灣營建業經濟動能。

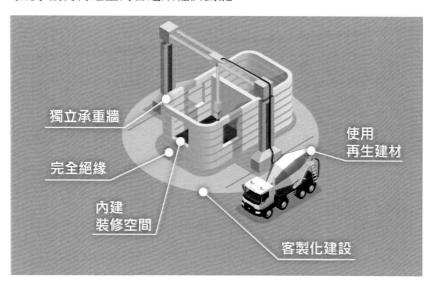

　　薪傳智庫數位科技股份有限公司是一家「定錨企業」（Anchor Firm）：指能誘發當地與該企業有關之供應商發展與創業的核心或主要企業，薪傳運用多元資源整合來落實，滾動式檢討未來推動 3D 列印房子發展政策，臺灣中小企業與新創產業的發展都需要政策的支持。從事營建業近 30 年經驗的我認為：此時有二點至關重要：(1) 營建產業相關營建法規與範疇能跟上國際腳步修正；(2) 倘若可獲得臺灣政府重視的中央與地方獎勵及補助措施會讓營建業更落實這國際化的長期策略；在智慧城市（Smart City）解決城市問題（遊民、人力、人口、成本），讓公民有購屋能力，更全面化解社會住宅供需不平衡。

（五）永續城市篇

永續城市未來都市

美濃區區長
鍾炳光

　　身為美濃區的大家長，打造美濃擁有更優質更美好的環境空間孜孜矻矻著每一個環節是所有美濃人的期望，更是我想要達成的目標。

　　根據聯合國糧農組織（FAO）統計，農田裡的作物品種在過去 100 年減少了 90%，我們吃的食物種類比過去祖先吃的更為單調。近年聯合國推廣「里山倡議」，希望把人與自然的距離重新拉近，有助於生物多樣性保育和永續發展。為了讓美濃區的農民在山腳下廣大平原生產糧食，許多傳統作物和農村特有文化，如：美濃的白玉蘿蔔、澄蜜香蕃茄等作物，是城市與農村共享淺山農業環境的果實；節能隔熱的建築改造，為減緩夏季高溫與氣候變遷勾勒出了具體的行動方案；街區的小塊綠地、屋頂和社區農園，也能一點一點串連成生態跳島，為都市裡的生物提供友善的棲息環境。為了讓其在這樣的環境裡延續，身為一區之長的我和區內的所有同仁一起努力爭取高雄市政府的資源，挹注重大建設在美濃區，諸如：美濃中正湖環湖環境設施工程、美濃文創中心、景觀橋、清水大排改善工程等等，維護城市完整的生態系，同時減緩氣候變遷帶來的風險，便是為生物多樣性保育付出行動，也為都市生活建構健康永續的未來，亦可提升美濃在地與友善環境的經濟發展，進而達到永續城市的發展目標。從永續的觀點來看，鄉村發展的問題與挑戰，可區分為經濟、社會與環境三大面向。缺乏在地就業機會，加速了農村人口外移、勞動力不足的惡性循環；農村文化被都市、全球文化覆蓋，失去了本地農村文化的魅力與多樣性；農業生產環境被過量的化學製劑破壞，生物多樣性為長期農業生產安全的必要條件，破壞帶來了世界糧食供應與食物安全的危機。綜合言之，農村是一兼具經濟、環境保育與文化活動等多面向的生活空間，為了確保人類生活條件的永續發展，如何解決農村地區的社會、經濟與環境的問題，是現下我們的當務之急。

　　所以美濃除了增加建設外，也加強社區營造、社福業務、獨居老人送餐服務，更是落實了減少貧窮，也辦理父、母親節表揚或是長青重陽節趣味活動，乃至民生基礎建設維護事項等，同時更為了我們偏遠鄉親的就醫轉乘或是洽公購物等，開辦免費庄頭巴士，都是為了減少不平等並戮力拉進城鄉在公共服務提供上的差距。

　　另外，為了吸引更多人到美濃遊玩，甚至深深愛上美濃，在觀光行銷方面，從彩繪大地花海季以及一連串的區政特色活動，使得遊客在漫遊美濃時有了不同的深度體驗；也讓本區近年除享有「微笑之鄉」美譽外，並在 2012 年榮獲臺灣十大觀光小城、參加 2018 臺灣城鎮品牌獎獲銅獎、入選 2019 年「小鎮漫遊年」台灣 30 經典／客庄小鎮等成果，皆是近年來區公所所有同仁在市政府的指導與支持，以及美濃公民社會等各個團體和民眾的共同努力及配合下的具體成果。

　　然而，身為一區之長以及在美濃土生土長的我來說，我並不會因為上述已經取得的區政成果而感到自滿，除了上述各項建設與施政工作的推動外，一個能留住人且吸引人的農村，必然需要一方面保留且對自己既有的文化感到驕傲與自信外，另一方面，也需要以更寬廣的心胸和開放的態度來尊重、接納和包容來自不同地方以及擁有不同文化的群體和人們。在這樣多元文化的觀點下，區公所全體同仁在市政府的支持下，也積極推動各項文化推廣與保存活動。例如，區公所與美濃各界（包括聖蹟會和廣善堂）在每年的正月初九會舉辦「迎聖蹟 · 字紙祭」的文化活動，藉以彰顯美濃傳統客家農村當中對於文字的珍惜以及對於探索浩瀚知識之海的尊重與渴望。另外，區公所也積極的和區內各個社團、地方仕紳與耆老共同努力，一齊努力保存和推廣美濃富含人文韻味的客家傳統文化，包括客家擂茶、粄條、油紙傘、客家藍染與藍衫衣服的製作等，也透過黃蝶祭的舉辦，藉以彰顯美濃區公所與各界對於環境保護的重視以及對於戮力保存各種不同物種生存環境的努力。

　　區公所除了努力的保存與推廣傳統客家文化外，區公所近來也和市政府、中央政府相關部會以及在中華民國徐生明棒球發展協會的共同合作下，已多年來共同辦理徐生明國際少棒錦標賽，透過該錦標賽的舉辦，希冀一方面能夠延續美濃人對於少棒精神與歷史的共同回憶，也希望透過在區公所與各界的共同努力下，將徐生明教練和棒球的運動家精神傳承給美濃的子弟，並提供美濃新生代的球員有一個與全國乃至於世界切磋球技與開拓眼界的機會。另外，也值得一提的是，區公所亦相當重視新移民在美濃的生活情形以及權益保障，1995 年時美濃即成為了臺灣各地方中開辦「外籍新娘識字班」的先驅，而在新移民於美濃安居樂業後，原本的他鄉已逐漸的成為了來自世界各地新移民的故鄉，以及安頓家庭和歸屬感的地方。而近年來，新移民在美濃地方經濟中所扮演的角色也愈發重要，其中，原本已逐漸沒落的水蓮（野蓮）產業，也在廣大的新移民群體的投入下不僅迎來了產業的復興、改善了許多新移民家庭的經濟狀況，也成為了美濃一個具有全國知名度且年產值破億元的在地以及友善環境的農產業代表。另外，區公所也利用美濃客家文物館的空間，並且和相關的在地新移民組織共同合作，一齊辦理新移民遷徙、移動意義與「家」的尋找等工作坊與活動，藉此協助新移民一方面透過定期的聚會消解思鄉之情，另一方面，也希望透過讓新移民與美濃在地社會有更多和更頻繁的互動，讓新移民能更好的適應在美濃的生活，並逐漸的在美濃落地生根。

綜上，我們期待因為這些深具在地美濃客庄文化特色的農產業田園活動，帶動觀光休閒人潮，促進經濟商機，加上配合在地人文、產業的特殊性，因地制宜、彈性整合運用各方資源，與時俱進，共同將美濃打造成更美好的永續城市，創造未來與願景。並讓美濃成為一個不僅是對環境友善、對傳統和多元與不同文化尊重與包容，也是一個同時能讓原有居民安居樂業，讓來自全世界新移民感到被尊重、被包容、被接納的一個宜居且生機盎然的富麗農村與風情小鎮。在此，身為一區之長的我，祈願美濃越來越好，所有居民安居樂業、平安健康。

（六）農業經濟篇

城市永續結合新農業創新

冬山鄉農會總幹事
黃志耀

　　根據聯合國統計，每年，全球所有食物生產中有 1/3（相當於 13 億噸、價值 1 萬億美元），在消費者和零售商的垃圾桶中腐爛，或由於運輸和收穫不當變質，這些食物在還沒進到你的胃之前，就先進了垃圾桶！與此同時，全球有 50 億肥胖人口，但仍有將近 10 億的人營養不良。所有國家都有權利進行經濟發展，但前提是必須尊重地球限度（參考：〈地球真的到極限了嗎？邁向永續發展刻不容緩〉）、確保永續生產與消費模式。在自由貿易的體系內，因為農產品市場價格的波動與氣候變遷的衝擊，讓許多農民與工人並未真正獲益，天還沒亮，許多農民已經展開勞動，但是全球有許多小農 1 個月的收入不到 1,200 元臺幣，沒有能力送小孩上學與吃飽，長期生活在恐懼與赤貧之中。新農業在已開發國家，解決生產端的浪費、廢棄物的產生以及普遍性的過度消費是首要任務，並協助開發中國家，強化技術及科技能力，發展以永續為目的的生產模式，提供公平且對環境友善之產品與服務，來鼓勵並刺激民眾消費，以滿足民眾的基本需求，進而提升生活品質與生產力，協助小農提升自信和能力外，「給他魚吃不如教他釣魚！」公平貿易以透明的管理方式與商業模式，建立一個可責信的供應鏈與商業網絡，讓消費者可以清楚追溯產品的來源，並鼓勵生產者使用永續的生產方式，產出對環境友善、對消費者更好更健康的產品，以夥伴的關係與生產者建立長期的合作，扶持生產者與農民提高產品品質，改善生產條件，培養他們獨立自主的能力，城市是全球永續戰役的關鍵節點，從農會出發擴散包容整個城市，實現環境共好的永續城市。

　　臺灣農業的無限可能就在冬山鄉農會開啟，循環設計改變地方創生，標榜全臺首座農業文創園區冬山 60 年老穀倉改建再出發。

　　我是冬山鄉農會總幹事，我要讓下一代勇於做夢，我們必須提供一個永續的環境，因產量少、產地偏遠、缺乏市場等資訊使單一的小農往往沒有議價權利，我們擴散提供生產者一種永續的方法來改善他們的生計。宜蘭冬山鄉農會改造 5 間有 60 年歷史的老穀倉；曾經是惡水的冬山河，經整治後已蛻變為「人與環境共存共榮」的希望之河；冬山鄉農會期許並賦予員工為冬山鄉產業發展有更多貢獻的使命，火車站前的閒置倉庫整建，經設計師巧手規劃，短短1 年的時間成就了「良食農創園區」的誕生，內部以玩 bar、買 bar、吃 bar、學 bar 等 4 項主題，各有不同特色，以農業、農民為根本，注入健康、創新、時尚等元素，打造出融合旅遊、體驗、美食、好禮、童玩的生活空間，為全臺第一個以農業為主題的文創園區「冬山良食農創園區」，園區有小農、青農等農特產品推廣區，標榜消費者可跟農友面對面縮短彼此距離，因良食農創園區鄰近冬山河、冬山火車站、老街，往上游可串聯仁山植物園、新舊寮瀑布、梅花湖，下游則有五十二甲溼地、冬山河親水公園，因此農會與旅行業者合作，希望以農業為出發點，規劃體驗遊程，用旅遊帶動地方經濟。農會也與雄獅等旅行社業者簽立合作意向，將結合農產、農作體驗，規劃系列遊程，盼以農業當火車頭，串連地方經濟，未來結合冬山鄉公所「冬山河之星」計畫。

良食
農創園區

—

EXCELLENT
FOOD AGRI.
PARK

（七）企業社會責任（CSR）報告書的策略模式與撰寫要領

企業社會責任（CSR）報告書的策略模式與撰寫要領

高苑科技大學商業暨管理學院教授
國家TTQS評核委員
行政院退輔會創業輔導顧問
李介綸博士

一、撰寫企業社會責任 (CSR) 報告的思維心法

（一）CSR 的定義

CSR，依據維基百科解釋是一種道德或意識形態理論，主要討論政府、股份有限公司、機構及個人是否有責任對社會做出貢獻。分為正面及負面：正面是指有責任參與；負面指有責任不參與，指企業超越道德、法律及公眾要求的標準，而進行商業活動時亦考慮到對各相關利益者造成的影響，CSR 的概念是基於商業運作必須符合可持續發展的想法，企業除了考慮自身的財政和經營狀況外，也要加入其對社會和自然環境所造成的影響的考量。

從 CSR 的角度來看，利害關係人主要是指股東、員工、消費者、環境、社區、競爭者、供應商和政府。企業對不同的利害關係人，應承擔不同的責任。20 世紀 60 年代，當時史丹福大學研究小組下的定義是企業存在目的不僅是為股東服務，還有許多關係企業生存的利害關係人，企業如果沒有他們的支持，就無法生存。在這情況下企業與相關利益者接觸時，試圖將社會及環境方面的考慮因素融為一體。因應企業的各利害關係人而編寫的企業社會責任報告書（Corporate Social Responsibility, CSR），也可稱為企業永續報告書（Corporate Sustainability Report, CSR），以報告書的方式，詳實揭露公司在永續經營及社會責任的目標、成果、承諾及規劃。

（二）CSR 的銓釋

節水節能、揭露綠足跡、淨灘、減碳減廢等環保措施，只是環保「基本功」，現在的環境永續做法，已經提升到如何將「綠色」納入營運中。CSR就是企業要「取之社會、用之社會」，不光只是替股東賺錢而已，還要對社會、環境的永續發展有所貢獻。例如，將回收資源變成新材料的「循環經濟」，就是近年的顯學，大愛感恩科技用回收寶特瓶來做衣服、遠東新用海廢垃圾來做球鞋，都是例子。並非只有製造業才能在環境永續上有所作為。零售業也可以在減塑、減廢、節能上，提出對環境友善的做法。例如荷蘭跟英國已經有無塑超市，家樂福也捐出即時品來減少食物浪費，並照顧弱勢。延續到今日，還是有很多企業常將社會責任等同於做公益，並以成立基金會、贊助藝文活動、認養公園、捐助弱勢族群等方式，推動社會責任。不過，這些外部的公益活動，雖是社會責任的一環，卻絕不是全部。

（三）CSR 與永續發展

聯合國永續發展鎖定目標有 17 指標包括：1. 消除各地一切形式的貧窮；2. 消除飢餓，達成糧食安全，改善營養及促進永續農業；3. 確保健康及促進各年齡層的福祉；4. 確保有教無類、公平以及高品質的教育，及提倡終身學習；5. 實現性別平等，並賦予婦女權力；6. 確保所有人都能享有水及衛生及其永續管理；7. 確保所有的人都可取得負擔得起、可靠的、永續的，及現代的能源；8. 促進包容且永續的經濟成長，達到全面且有生產力的就業，讓每一個人都有一份好工作；9. 建立具有韌性的基礎建設，促進包容且永續的工業，並加速創新；10. 減少國內及國家間不平等；11. 促使城市與人類居住具包容、安全、韌性及永續性；12. 確保永續消費及生產模式；13. 採取緊急措施以因應氣候變遷及其影響；14. 保育及永續利用海洋與海洋資源，以確保永續發展；15. 保護、維護及促進領地生態系統的永續使用，永續的管理森林，對抗沙漠化，終止及逆轉土地劣化，並遏止生物多樣性的喪失；16. 促進和平且包容的社會，以落實永續發展；提供司法管道給所有人；在所有階層建立有效的、負責的且包容的制度；17. 強化永續發展執行方法及活化永續發展全球夥伴關係。原先勞動部的投標案最早就把 CSR 納入審查指標，包括薪資水準、調薪幅度與生活平衡措施，爾後增列職安項目並擴及其他部會。目前經濟部則推動社會組織創新也是鼓勵企業或組織朝此目標邁進。

然而 CSR 要能成為企業的 DNA，就必須由上到下來推動。企業最高層必須重視且參與，才有可能動員員工。否則，基層員工往往忙於各自工作，往往只認為 CSR 是人資或公關部門的事。因此唯有最高管理階層真正展現出對 CSR 的支持與承諾，才有可能讓員工都認真看待 CSR。有些企業已經將 CSR 列為績效考核的 KPI 之一。例如克蘭詩的各國分公司總經理的年度績效考核，有 10% 來自於 CSR 的表現；聯合利華則規定旗下每個品牌，每年至少要推動一項永續專案。

二、編寫 CSR 報告的策略運作模式

（一）規劃 CSR 的策略三個層次

發展至今，CSR 策略大致可分為三個層次：第一層次是企業參與慈善和捐助活動，但與企業核心業務沒有直接關係；第二層次是透過加強管理和生產方式的改進，減少對社會和環境的負面衝擊，此時已將外部因素導入經營決策中；第三層次是兼顧經濟、社會和環境的共同利益，追求利潤和社會責任並重。許多公司導入 CSR 方案的績效不如預期，往往是因為在著手規劃 CSR 的具體策略前，並沒有先清楚地思考何謂企業的「社會責任」，沒有找出 CSR 的適當主題和落實方式，也沒有將 CSR 的觀念導入企業經營的視野中。

（二）CSR 的策略執行模式

從企業的角度看來，以 CSR 的策略執行模式而言，與企業發展策略的密切程度關係，大致分為以下五種：

1. 資源協助策略模式：

企業的執行方式包括經費補助、不指定特定用途的慈善捐助、經費贊助，以及企業支援提供一般性支援協助等。這些資源協助，多半以「單次」或「短期式的經費」，及一般人力協助為主。這樣的投入模式亦與企業本身經營業務專業不見得有關聯，CSR 與公司使命及營運兩者獨立，甚至有些支持是基於企業主的個人意志而來，因此，若企業易手、換人接班，這些支持便可能中斷或產生變化。由於與企業經營專業無涉，經

營階層容易產生「施捨」與「不樂之捐」的偏頗心態,對於建立夥伴關係來說,挑戰也更多。在臺灣這是目前佔比最大的協助模式,當中又以經費贊助模式最多。地方創生案例如台積電資助前美國大使館整修等。

2. **專業協助策略模式:**

此模式的 CSR 執行方式包括企業職員提供專業技術的志願協助、成功經驗分享、協助能力建構等。因為能與公司的業務專長領域相結合,所以能夠提供更進一步的互動合作關係。此模式建構在本業上,會讓員工更願意投入,後續也可能回饋到企業業務。著名案例如東鋼藝術家駐廠創作專案:每年,東鋼邀請一位國內藝術家和一位外國藝術家在苗栗工廠工作,在居住的幾個月中,東鋼提供廢鋼、專業機器和技術人員等資源。完成創作後,東鋼會辦理發表招待會來展示其作品。從辦理的經驗中,也讓東鋼的員工對藝術創作產生興趣。

3. **資源媒合策略模式:**

資源媒合模式則由企業擔任平台角色,規劃想法,以訴求爭取聚合各方企業 CSR 資源共同投入,針對不同課題解決需求,也媒合不同專長或資源擁有者共同投入。相關案例如信義房屋的「社區一家、全民社造行動計畫」,從第一階段(2004-2008)的申請案多以延續與宣傳地方產業文化開始,到第二階段(2009-2014)青年返鄉提案,成為傳承文化與永續發展的新興力量;再來發現許多社區需要提案方法的培力協助,否則無法在媒合平台上成案。2015 年則更名為「全民社造行動計畫」調整為「學習互動參與」。2020 年更呼應教育部 108 課綱「互動、共好」探索社區問題,改造自己家園。

4. **商務合作策略模式:**

此模式的執行方式包括社會採購,帶動消費者購買具有社會使命的產品和服務,強調額外的影響力。或是與合作夥伴提供客製化或優惠或免費的產品,共同專案,計畫執行等等。從商業模式角度構思 CSR 投入策略,因此合作方案可能是共通業務開發或專案合作,也可能是顧客商務關係,卻不是追求理論極大化的一般商業關係。

5. 策略夥伴策略模式：

這種模式的 CSR 執行方式包括共同辦公室，或是事業策略的融合。而企業與 CSR 合作夥伴間屬於高度整合關係，甚至整合進集團中，思考透過 CSR 執行來達成社會目的，集團下的部門也會透過 CSR 執行，為公司永續成長製造動力，在顧客選擇與關係經營中，兼具商業獲利與社會責任的思維。

（三）CSR 的策略衡量績效

由於社會和環境議題對企業經營的影響日增，尤其是機構投資人在選取投資標的時，除了考量傳統的財務績效，也會將環境績效與社會績效一起列入考量，因此企業也要對外揭露非財務績效的永續發展報告書，是目前國際發展的主流。各方開始出現將報告書標準化的呼聲，其中最具影響力的機構，是由聯合國環境規劃署（UNEP）與「對環境負責任之經濟體聯盟」（CERES）在 1997 年共同成立的「全球永續發展報告書協會」（Global Reporting Initiative, GRI）。GRI 報告是全球唯一由多方利害關係人，包括企業團體、勞工、人權、會計、環境及投資機構等組織，所共同制定的報告書架構，主要提供一套廣為大眾接受的體制，供組織報告其經濟、環境及社會績效，架構包含永續發展報告指南、各類指標規章、技術規章及行業補充指引。

隨著企業全球化時代來臨，愈來愈多人開始察覺到跨國企業在永續發展及社會責任議題上的表現。這個助力可視為一種催化劑，迫使企業去定義自己獨特創舉以邁向永續發展，並滿足外界對企業善盡 CSR 的期待。企業詳細及明確地揭露非財務績效資訊，將使企業有機會向社會大眾承諾與企業經營及發展願景有關的具體目標。

三、實踐 CSR 報告書的編製執行步驟

CSR 是以企業之利害相關人為主體，旨在企業對於利害相關者關切議題之回應與責任履行，以表達企業之當責性；而企業永續則是以企業為主體，旨在考量環境與社會面向之經營策略，提升企業競爭力。不論是 CSR 報告書或企業永續報告書，皆是一種企業與利害相關者進行溝通之工具，考量企業在經濟、環境、社會與文化中每一個面向的運作，制定策略並藉由透明與正當的員

工發展方式使企業長期經營。一份適當的 CSR 報告書或永續報告書，應充分並透明的揭露資訊。

在釐清上述策略思維後，企業在準備製作及發行相關報告書時，一般有以下四大步驟：

（一） 第一步驟：實施 CSR 之制度架構、政策與行動方案以確認申報方式：

首先是報告書製作的目的是什麼？瞭解閱讀對象在哪裡？然後清楚瞭解為什麼企業必須要申報在 CSR 面的執行績效？該依循什麼樣的報告標準？並應該申報什麼樣的 CSR 資訊？如何定義及蒐集 CSR 有關的資訊？以及依據什麼樣的原則來申報 CSR 的資訊及數據？最後定性的特徵是什麼（關聯性、真實性及透明度……等），因為閱讀對象不同就有不同解讀：消費者看 CSR 報告書，看出產品好安心；大學生看 CSR 報告書，看出企業好新奇；員工看 CSR 報告書，看出職涯好福利；股東看 CSR 報告書，看出投資好布局；研究者看 CSR 報告書，看出趨勢好分析；NGO 看 CSR 報告書，看出合作好關係。

（二） 第二步驟：闡明主要利害關係人及其關注之主題與主題邊界 CSR 報告書的內容規劃：

先明白報告書所要傳達的訊息是什麼？如何定義與 CSR 有關的案例？接者清楚報告針對的讀者群在哪裡？以及哪些類型的利益相關者應該納入考量？又該如何納入？然後想要申報什麼樣的 CSR 資訊？可以從管理程序中獲得什麼資訊？內部組織應該如何改善，才能將所謂重要的 CSR 議題、活動及指標確實的傳達出去？與可以從那些系統取得重要的資訊？如何查證資訊的正確性？

企業從過去大多採用短期的、緩慢增加的 CSR 專案，到目前越來越多企業開始採用更為全面、且具有社會使命的策略型營運模式。其他的轉變還包括：把 CSR 與商業模式整合；從 CSR 策略轉型成具社會目的的商業模式；採用數據資料做為社會目的策略規劃的基準；積極地衡量、追蹤與揭露企業對社會的影響。企業從過去把 CSR 視為公司需承擔的社會責任，到現在把 CSR 視為營運公司的方式；企業對於 CSR 的想法也從過去的加分項目，轉變成為營運的成功關鍵。

（三） 第三步驟：公司於推動公司治理、發展永續環境及維護社會公益之執行績效與檢討：

　　這個報告製作的步驟大部分的時間多花在草擬報告及最終報告的美編製作上。這個階段要注意的最大的問題是「數據蒐集期間的定義」。由於報告書的數據是以一整年為蒐集期間，報告數據絕大部分都在製作初期取得，但是整合與分析的作業通常是在最後階段才執行，有些企業可能會誤用後期的數據，做為報告書績效展現的基準，往往扭曲了數據展現的真實性。因此，企業如何在這個階段察覺數據分析所帶來的影響，是相當重要的。此外，有些較有報告書製作經驗的企業，已開始引入第三者確證（third-party assurance），以提升報告書的可信度。重點是，企業如果願意花費更多的時間在資訊分析的作業上，相信製作出來的報告書品質也會相對提高。

（四） 第四步驟：報告書的發行、發送與使用後檢討改善：

　　針對先前所設定的讀者群與利害關係人進行資訊散布，對象包括企業內部員工及外部重要利害關係人、NGOs 以及金融機構等，最後一個重要的步驟，那就是做為檢討與學習未來之改進目標與管理方針方向。在 CSR 報告出版後，企業應該設回饋機制，主動蒐集社會大眾對報告的回饋資訊，讓不同的利害關係人針對該企業的 CSR 績效給予回應。進行改善報告書內容包括長程與短期展望的討論等，並將心得納入報告書的製作策略中，做為下一版報告書的參考。

五、持續改善報告的超前部署

（一） 可遵循的 5 項關鍵元素

　　近 30 年來致力倡議企業永續的國際著名非營利組織「企業社會責任協會」（BSR）執行長 Cramer 曾撰文指出，面對未來全球利害關係人對企業有著更多的期待，企業可遵循 5 項關鍵元素，出版更具溝通效益的永續報告書。

1. **第一關鍵元素**：永續報告編纂現在正進入一個重要的轉變階段，這項轉變極可能會決定未來十數年的報告揭露與撰寫方向。這項此刻出現的轉變，有其重要的意義，並且直接凸顯出版報告書的核心命題：為何企業必須出版報告書？

2. **第二關鍵元素**：企業並非純粹為出版報告書而出版報告書。企業出版永續報告書的真正目的，是激勵企業作為，加速其永續轉型。更重要的是，企業的永續報告也能成為投資者、公民組織以及政府的優秀參考指標，同時為永續與平等的未來，提供更好的評斷依據。

3. **第三關鍵元素**：幫助商業轉型及加強企業績效。藉由依循永續報告架構，幫助企業發展具有更高韌性的商業策略，以吸引更多投資並強化公司營運作為。

4. **第四關鍵元素**：為社會帶來更好的永續發展。輔導企業增設專職永續委員會與重大議題管理方針，幫助企業達成聯合國永續發展目標、《巴黎氣候協定》，以及其他國際人權指標。

5. **第五關鍵元素**：加速平等與永續經濟的決策。透過報告書，企業應能快速提供投資人、消費者、公民組織與政府等，各方利害關係人必要的資訊，以加速激發雙方永續發展的步伐。只有企業確實地依循與實踐這些永續報告的框架與準則，上述目的才可能被真正實踐。BSR 透過累積近30 年在永續業界的經驗，觀察企業與倡議組織的作為，以及是否尚待完成之處，希望能夠歸納出一個新觀點，指引未來企業永續報告書，應該如何有效溝通。

（二）時代脈動 CSR 創新與環境成本會計系統的建立

企業經營隨著時代的脈動，在不同階段會面臨不同的挑戰與創新，1970年代為成本降低年代，1980 年代為品質年代，而在講求服務的 1990 年代，在目前的環境下，企業想要掌握致勝的關鍵，除了應該瞭解並滿足顧客多變的口味外，更必須具備整合及創新的能力，以適應目前這個多元化競爭的經營環境。

國際環境管理標章 ISO14000 在 1996 年發布，強調社會必須透過產品生命週期的評估及環保規章的限制，將以往外部化由社會負擔的環境成本轉為內部化，自此，環境成本會計系統也應運而生。也就是說，為了達成生態效益，邁向企業永續經營；企業必須從產品設計、製程改善、污染防治、售後服務、至廢棄物回收處理等整個生命週期，積極進行改善。企業永續發展的目的在於，確保組織於營利的同時，考量社會的環境的成本。換言之，其核心精神在於如何將改進社會和環境的力量，轉化為業績的提升。為達此目的，發展健全的企業整體策略為第一要務。管理階層人員和專業人士如會計師可以顯著的改變組織運作的方式，但是革新性策略的推行與否，治理階層的遠見和領導能力為最重要的催化劑。治理階層之責任在於建立和設定組織目標。

簡言之，企業必須協同股東、員工與供應商從產品設計、原料選定、製程改善、污染防治、售後服務至廢棄物回收處理等整個生命週期，進行積極的改善。在過程中除了社會與環境因素，財務因素亦十分重要。三者涉及產品或服務在其整個生命週期的成本、考慮營運成本和處理成本，以及收購成本。因此，環境成本會計被定位為促進企業永續價值達成的重要技術。以達成：1. 顯現出對永續發展的承諾；2. 幫助以確保永續發展是嵌入在一個組織活動內；3. 示範一個組織如何執行，展現組織的承諾並發展量化與具時限的永續發展目標。

（八）CSR 實務商模戰略──借鏡 B 型企業模式

CSR實務商模戰略──借鏡B型企業模式

台灣影響力投資協會理事長
張大為

　　CSR 是現今國際企業的一門顯學，而且也是跨國企業選擇產業供應鏈夥伴的一個重要條件，從而延伸出 ESG 的評鑑模式。其實包括早期的 CSR、到後來的社會企業、公司治理、赤道原則（Eps-Equator Principles），碳足跡與綠色金融，也都是 CSR 不斷演進的過程產生的一些新原則。CSR 在臺灣上市公司及部分非公發的大型企業已經做得很好，一部分產業上市公司被要求每年要公開 CSR 報告，雖然因產業特性不同沒有絕對固定的格式，不過各揭露公司的報告也大同小異。只是 CSR 主要還是針對大型企業所設計的評鑑規範，製作成本高昂，因此即使上市櫃公司一體適用，但仍然不適合臺灣廣大的中小企業的評鑑，更不用說微型企業完全沒有條件去做 CSR 報告。但是就 B 型企業來說，則是另一種 CSR 的評鑑機制，且可以完全適用於大型企業和中小企業，甚至微型企業只要有意願，也可以做得到 B 型企業的評鑑要求的項目。我不會說 B 型企業是改良式的 CSR，或者是簡易式的 CSR，因為它完全是一個獨立的評鑑標準，但是完全符合 CSR 的大原則要求。基本上申請並不困難，也不複雜（可能每個企業主對複雜的定義不同），這些要求只是要申請的企業去身體力行，並且每 3 年重新檢視，以確認自己沒有離開 B 型企業的軌道，讓企業可以符合 B 型企業永續經營的精神。因此 B 型企業強調企業在日常營運中很自然的養成習慣去做一些事，並且利用做這些事的過程創造一個營利的新的商業或營運模式來獲利。

　　以下我以個人過去經營中華徵信所的經歷，以及後來成為 B 型企業共同創辦人的過程，並帶領中華徵信所成為 B 型企業的經驗，和關心 CSR 發展的企業及讀者分享我的經驗。首先我必須說，跨進 B 型企業，我並非跳躍式的進化，而是長期以來我就在思考一家好的企業需要什麼主客觀條件？因為我長期在中華徵信所擔任總經理，公司每天都在為企業進行徵信調查和企業評分，其中除了總體經濟層面、產業的週期外，舉凡公司創立時間長短、公司背景、營運狀況、財務狀況、企業負責人的票信、正負面新聞，是否有申請專利、上游客戶下游供應商、關係企業情況等等，都在徵信報告的評分項目之內，但我總覺得還缺少了什麼因素？正巧在 20 年前有機緣開始接觸 CSR，我才以一家企業經營者的身分親身瞭解到，一家好的企業除了要能營利之外，透過什麼樣的模式營利也很重要。同時企業營運取之於社會也要用之於社會，因此應該負擔「企業社會責任」。縱使一家企業能夠賺錢，但是以什麼樣的模式獲取利潤，也是非常重要的。（舉例來說，若一家企業晚上違法傾倒廢料排水來降低成本賺取黑心錢，再用這些錢做慈善公益來贖罪，它終究還是一家違法經營的企業。）那時我們和喜瑪拉雅基金會一起研究「企業社會責任」的評鑑架構。記得當時還有澳洲的 CSR 研究機構來臺訪問，告訴我們澳洲已經推出設定一些評鑑機制來包裝 CSR 的 ETF 組合，並且研究發現納入組合的成份股都是獲利報酬很高的企業，我從中意識到 CSR 和企業經營的連結正向重要性。我當時很想要把 CSR 評鑑納入我們徵信報告評分的一個項目，但那時臺灣公司治理尚未十分成熟，CSR 的環境才剛起步，且 CSR 評鑑項目複雜，如果要放在一份徵信報告中，製作報告的時間和報告的成本都非常昂貴，因而引進到徵信報告的計畫被擱置，但是我們把企業環保因素罰款的項目加到徵信報告中，成為報告中的一項內容。同時我並沒有放棄追求 CSR 的想法。在公司裡由我去發起推動「地球環保日」，讓同仁自發性的在假日和公司一起去做環保公益，在公司內舉辦二手商品義賣，把義賣所得捐給慈善機構。在國內發生重大災難時，鼓勵同仁捐出一日所得和公司一起提供捐款給受災戶。可是我覺得只是單純做公益還是不夠。因此 2014 年又去接觸「社會企業」，瞭解「社會企業」的成立目的，是為了要解決某一個社會問題。而在解決這個社會問題的同時，是透過設計一種服務或產品的過程收費取得營業收入，這筆營業收入就可以用來處理要解決的社會問題。我們辦了一些社會企業的講座論壇，把經營者請來和我們的客戶及金融業者分享他們的發展和成功經驗，目的在推廣社會企業概念。

看起來社會企業和現今的影響力投資潮流有點類似，但是社會企業不是營利為主要目的，和影響力投資還是完全不同的概念。在接觸社會企業的過程，2015 年才真正接觸到「B 型企業」。那麼究竟什麼是 B 型企業呢？簡而言之，就是在環境、員工、客戶、社區及公司治理五大面向，能夠達到一定的標準，並通過美國 B 型實驗室（B Lab）的認證後，就可以成為 B 型企業的一員。換言之，B 型企業並不是標榜社會企業，而是在於標榜一家企業不但要能夠營利，而且在這個經營的過程中，秉持企業的道德良知，在自己的能力範圍內去塑造一個平和的、善意的、公益的企業與社會環境。

從 2007 年 B Lab 發出首張 B 型企業認證至今，全球已經有超過 3,600 家的 B 型企業，但是佔全球企業總家數的比重仍然鳳毛麟角，而臺灣現有 29 家 B 型企業，數量居於亞洲各國前茅，但是在臺灣超過 162 萬家大小企業及商號中比例尚還不及 0.002%，顯示這樣的一個「好企業」的理念，需要有強大的力量去推動。因此如何把「世界上的 B 型企業帶到國內」，就成為我與連庭凱、胡德琦、陳一強三位夥伴在 2016 年共同創辦 B 型企業協會的動力。我們透過自己成為 B 型企業的經歷，來影響、帶動其他的企業跟進。即使沒有主動申請 B 型企業認證，也仍然可以成為一家符合 B 型企業的好公司。前面敘述過「B 型企業」強調的是企業要身體力行。所以必要在環境、員工、客戶、社區及公司治理五大面向有深度的著墨。透過 B Lab 官網去填寫問卷，檢視企業本身原本就做了哪些符合 B 型企業的指標，不但是自身感覺到有做，而且要能拿出實際的資料證明本身有做到。對於在五大面向沒有做到的項目，就要去思考如何能夠改善以符合認證的要求。不但企業本身要有行動力去實現，還要感召企業員工願意一起配合去做，並且養成日常習慣。

猶記當時我們在申請 B 型企業時，就在看我們本來如何節約能源？在中午休息時間有隨時關燈的習慣，夏天冷氣維持在 28 度，辦公區的照明用的是 LED 燈，老舊的燈管更換一律採用 LED 燈管。並用每次的水電費單來檢視我們節約能源的成果。對我們來說無紙化作業並不容易，但是我們原本就朝向這個方向去改善，把電腦化流程設計得更深化，各種調查的資料審查都改為電子審查。雖然沒辦法做到百分之百的無紙化作業，透過事務機器機的管理，我們可以瞭解哪些部門或哪些同仁的用紙量顯著減少，這都是身體力行的證據，在申型企業時可以提出的書面證明。對於在公司內中午用餐完全杜絕免洗筷、免洗碗進入公司，要求同仁自備環保碗筷，甚至設計環保筷及吸管送給客戶，希望以實際行動影響客戶和我們一起做環保，都可以為申請 B 型企業加分。在照顧員工方面，我們嚴禁員工加班，希望員工以效率代替加班，能夠準時下班，也與家人享受週末的休息。並且我們連續 3 年主動調薪，並且我主動開始設計員工分紅的制度。並且我願意把公司的盈餘拿出來提供給員工們比過去更高的年終獎金，在我 2017 年卸任總經理之前，連續 4 年我提供員工與過往相較最高的年終獎金。這並不是因為要通過申請 B 型企業而做，而是 B 型企業讓我領悟到員工是可貴的資產，在我長達 30 年的總經理生涯，我們幾乎沒有一年虧過錢，也沒有一年沒發年終獎金，可是在 B 型企業的精神下，我更願意與員工分享公司經營的成果，我希望一起打拚的同仁認同我的理念，認同公司這個大家庭。我體會到做為一個幸福企業的喜樂。這些都是我在經營企業的過程裡身體力行的經驗。

當然 B 型企業的第 1 項條件就是企業要有營利。並且用不同的經營模式去經營企業。不是用降低成本來擠出獲利（我並沒有說降低成本是不對的，而是強調不要依賴降低成本來成就獲利，因為降低成本是本來就該做的事），所以力求作業的自動化，以自動化的模式來節省徵信報告的作業時間，包括自動化模組、自動化的資料比對、除錯，自動化的預測以及簡易自動化報告。在這方面我們轉型得非常快而且好。在資料庫運用上，我們把資料切割得更精緻但也更容易查詢，我們因而改變了客戶的基礎，把視野放在國際客戶上，因此我們不需要刻意降低成本來產生獲利，而是因為我們改變了作業模式反而提升了獲利。我發現當我們願意開始做的那一刻起，我們就和員工一起改變公司，而且停不下來。更令我感到欣慰的是，我們已經不需要透過評鑑來證明自己更好，因為我們自己就會產生想要更好的改變念頭。也透過 B 型企業協會打造 B 型企業生態圈，我們的採購儘量透過同是 B 型企業的夥伴公司，同時也要求供應商和我們一起做，以擴大 B 型企業的影響力。那時包括嘉威會計師事務所、銘宇興業、嘉澎塑膠等十家堪稱臺灣 B 型企業先驅的企業，取得 B 型企業認證之後對內部員工、公司治理方面的改變為何？對外部的客戶、供應商，甚至是對社會、對環境都有正面的影響。我們的共同點是都是中小企業，同樣願意以「不畏艱辛」、甚至「唐吉訶德般勇於做看似不可能的夢」的毅力和精神，去挑戰 B 型企業的認證，重新詮釋企業不一樣的經營視野和未來。

企業反思的覺醒由「綠色企業」、「社會企業」、「永續經營企業」，一路演變到「B 型企業」的誕生，但在態度上和行動上「B 型企業」是比起前幾種企業型態要來得更積極的一種企業類型，以新的策略來達到企業生存的目的，反而是越能賺錢的企業！原因在於，改變了經營想法，勇於承擔企業的責任，敢於做他人所不敢嘗試的事，因而整個商業模式也跟著改變。我相信企業無論規模大小，只要有心去實踐，每家企業都可以成為 B 型企業，即使不能通過 B 型企業認證，但是只要願意多承擔一點社會責任、多伸出一雙善心的援手，多存一點社會善念，就可以讓企業本身變得更好，也可以讓整個社會氛圍多出一分美好。B 型企業塑造的是現今和未來企業經營的典範，在經營理念上以「誠」為貴，以「將心比心」為出發。我個人一路走來的感受是，B 型企業是中小企業最容易實現的 CSR 縮影。所以當臺灣想要打造 CSR 社會理念以及要融入轉化成企業商業模式時，借重 B 型企業的模式，創造臺灣中小企業 CSR 評鑑模式將是當務之急。同時我也呼籲中小企業要自己動起來，無論參與哪一種 CSR 評鑑模式，自己要有身體力行的覺醒，我們相信只要開始做，企業就會改變而且變得更好。

參考文獻

- 文化發展與企業社會責任網（2020 年 6 月 19 日）。CSR for Culture—企業與藝文的協力模式。取自 https://csr.taicca.tw/csr-for-culture-%E4%BC%81%E6%A5%AD%E8%88%87%E8%97%9D%E6%96%87%E7%9A%84%E5%8D%94%9B%E6%A8%A1%E5%BC%8F/
- 王君瑭（2020 年 12 月 10 日）。最強館長棄百萬年薪！師千萬退休金只為做「這件事」！取自 https://www.businesstoday.com.tw/article/category/183035/post/202012100011/%E6%9C%80%E5%BC% B7%E9%A4%A8%E9%95%B7%E6%A3%84%E9%99%BE%E8%90%AC%E5%B9%B4%E8%96%AA%EF%BC%81%20%E7%A0%B8%E5%8D%83%E8%90%AC%E9%80%80%E4%BC%91%E9%87%91%E5%8F%AA%E7%82%BA%E5%81%9A%E3%80%8C%E9%80%99%E4%BB%B6%E4%BA%8B%E3%80%8D%EF%BC%81
- 中部科學園區管理局永續經營社會責任網站。如何撰寫 CSR 報告書？取自 http://www.ctspcsr.com.tw/article_list/view_article_detail/?id=41
- 行政院（2019 年 1 月 30 日）。循環經濟推動方案。取自 https://www.ey.gov.tw/Page/5A8A0CB5B41DA11E/18ef26a4-5d05-4fb3-963e-6b228e713576
- 行政院（2020 年 3 月 11 日）。前瞻基礎建設計畫—奠定未來 30 年國家發展根基。取自 https://www.ey.gov.tw/Page/5A8A0CB5B41DA11E/9cf2eef1-e2d2-4f37-ba6e-9498deb422b4
- 吳姿賢（2020 年 10 月 21 日）。每平方公里 21 萬件垃圾！台灣西海岸 8 大髒點，海底垃圾密度是全球 1.5 倍。取自 https://www.businesstoday.com.tw/article/category/80392/post/202010210015/%E6%AF%8F%E5%B9%B3%E6%96%B9%-E5%85%AC%E9%87%8C21%E8%90%AC%E4%BB%B6%E5%9E%83%E5%9C%BE%EF%BC%81%E5%8F%B0%E7%81%A3%E8%A5%BF%E6%B5%B7%E5%B2%B88%E5%A4%A7%E9%AB%92%E9%BB%9E%EF%B-C%8C%E6%B5%B7%E5%BA%95%E5%9E%83%E5%9C%BE%E5%AF%86%E5%BA%A6%E6%98%AF%E5%85%A8%E7%90%831.5%E5%80%8D
- 吳雁婷（2015 年 12 月 20 日）。世界上最遙遠的距離，是我們吃不完的食物進了垃圾桶，卻送不到那些需要的人手中。取自 https://www.seinsights.asia/article/3289/3271/3748
- 林佩萱（2017 年 8 月 18 日）。搞懂 CSR 關鍵 20 問一次解答。取自 https://www.gvm.com.tw/article/39488
- 林佩萱（2017 年 9 月 29 日）。台灣 20 家 B 型企業獲認證 榮登亞洲第一。取自 https://www.gvm.com.tw/article/40287
- 林務局（2020 年 4 月 29 日）。推動臺灣里山倡議的策略架構。取自 https://conservation.forest.gov.tw/0002057
- 威浦斯科技。全外氣自然空調機。取自 http://www.we-plus.com.tw/productsType.php?ProductsTypeID=05
- 美濃農村田野學會。取自 https://meinungfield.wordpress.com/%E9%97%9C%E6%96%BC/
- 陳姿蓉、吳宜靜（2020 年 6 月 8 日）。2020 世界海洋日：《淨灘手冊》讓海乾淨 向海致敬。取自 https://e-info.org.tw/node/224927
- 張亦惠（2021 年 1 月 7 日）。花光退休金 免費課輔連辦 10 年。取自 https://www.chinatimes.com/newspapers/20210107000596-260107/?chdtv
- 葉凱平（2018 年 6 月 25 日）。只要親自淨灘過一次。取自 https://www.greenpeace.org/taiwan/update/1765/%E5%8F%AA%E8%A6%81%E8%A6%AA%E8%87%AA%E6%B7%A8%E7%81%98%E9%81%8E%E4%B8%80%E6%AC%A1/
- 鄭君平（2008）。企業社會責任入門手冊。臺北：天下文化。
- 臺南市政府農業局（2014 年 11 月 5 日）。農會結盟一起大步向前行。取自 https://agron.tainan.gov.tw/News_Content.aspx?n=1262&s=150588
- 嘉威聯合會計事務所。B 型企業運動，讓商業成為改變世界的力量。取自 https://www.jwcpas.com.tw/publications-Book3-show2.php?book3p_id=236
- 維基百科（2020 年 9 月 12 日）。水域生態系統。取自 https://zh.wikipedia.org/wiki/%E6%B0%B4%E5%9F%9F%E7%94%9F%E6%80%81%E7%B3%BB%E7%B5%B1
- 維基百科（2021 年 3 月 13 日）。企業社會責任。取自 https://zh.wikipedia.org/wiki/%E4%BC%81%E6%A5%AD%E7%A4%BE%E6%9C%83%E8%B2%AC%E4%BB%BB
- 廖靜蕙（2016 年 12 月 19 日）。原來務農收入這麼好！美濃農會帶頭實踐「里山精神」。取自 https://www.thenewslens.com/article/56889
- 張議晨（2021 年 1 月 16 日）。標榜全台首座農業文創園區，冬山 60 年老穀倉改建再出發。取自 https://udn.com/news/story/7328/5179012
- 臺灣證券交易所公司治理中心（2021 年 2 月 3 日）。企業社會責任簡介。取自 https://cgc.twse.com.tw/front/responsibility
- 魯皓平（2018 年 12 月 18 日）。大海正在窒息！全球海漂塑膠垃圾 80% 來自亞洲。取自 https://www.gvm.com.tw/article/55332
- 聯合報系願景工程（2020 年 06 月 09 日）。「如果，海有你」專題報導。取自 https://today.line.me/tw/v2/article/QnKjr0
- 顏和正（2019 年 1 月 3 日）。什麼是企業社會責任？一次搞懂關鍵字 CSR、ESG、SDGs。取自 https://csr.cw.com.tw/article/40743
- Aron Cramer & Dunstan Allison-Hope（作者）；CSRone 施奕丞（編譯）（2020 年 12 月 1 日）。BSR：未來永續報告書 5

大關鍵元素。取自 https://csrone.com/topics/6569

- Bowen, H.R. (1953). *Social Responsibilities of the Businessman*. New York: Harper & Brothers.
- CSRone 永續報告平台、政治大學商學院信義書院、資誠聯合會計事務所（2017 年 3 月）。2017 年台灣永續報告現況與趨勢。臺北：CSRone 永續報告平台、政治大學商學院信義書院、資誠聯合會計事務所共同出版。
- European Union (EU). (2021, Feburary 3) A renewed EU strategy 2011-14 for corporate social responsibility. Retrieved from https://eur-lex.europa.eu/legal-content/EN/TXT/?uri=CELEX:52011DC0681
- Global Reporting Initiative (GRI). 2016. Sustainability Reporting guidelines: Version 4.0. Amsterdam: GRI.
- Jessica Seddon et al.（作者）；CSRone Hope Wang（編譯）（2019 年 7 月 9 日）。隱形殺手！空氣污染沒你想的那麼簡單。取自 https://csrone.com/topics/5631
- KPMG（2021 年 1 月 12 日）。2021 玉山安侯論壇：CSR 大不同 社會創新是新寵。取自 https://home.kpmg/tw/zh/home/media/press-releases/2021/01/tw-csr-social-innovation-seminar-2021.html
- Levitt, T. (1958). The Dangers of Social Responsibility. *Harvard Business Review*. 36, 41-50.
- PChome 個人新聞台（2020 年 6 月 7 日）。海洋是生命之母。取自 https://mypaperapp.m.pchome.com.tw/carled31599/post/1380326363
- Searcy, C. & Buslovich, R. (2014). Corporate Perspectives on the Development and Use of Sustainability Reports. *Journal of Business Ethics*. 121: 149-169.

薪傳智庫數位科技是協助企業匯聚 CSR 實務與將 GRI 準則
短中長期融入企業日常活動與經營策略的全球好夥伴。

Content

Foreword

This is a developing topicand it's becoming increasingly important:

The 17 Sustainable Development Goals set by the UN have not reached its end. The 2030 Agenda for Sustainable Development in response to COVID-19 is like the social contract of the new world, corresponding to The Social Contract by Jean-Jacques Rousseau in the 18th century. The definition of a free man has expanded to the digital world and the needs of all parties, developing a frame of imagination for SDG18, such as digital collaboration, outer space, the Laos Target, etc.

The SDG18 on digital collaboration acknowledges the threats on individual rights, social equality, and democracy, and these risks and uncertainty are magnified by "digital divide". Corporate social responsibility based on digital technology and deployed on the spirit of ESG and the GRI Standards has a chance to become a financial indicator as important as EPS in the near future. Therefore, the topic of SRI becomes more mainstream, which happens to serve the purpose of private sector enterprises to maximize profit, meanwhile makes enterprises pay attention to the social responsibility they should take. It is a practical strategy beneficial to human sustainability.

CSR development is a process of feeling its way forward. There are few practical local teaching materials in Taiwan, most of which are course briefing in development. Today, I am glad to see the publishing of *TransAct's CSR Practical Business Model Strategy*. It represents that there are already roots planted in the area of CSR in Taiwan, and they are embedded in the corporate DNA of pursuing sustainability development. Such reflexive development corporate activity indicates that the four pillars of civic responsibility (society, economy, environment, culture) which European welfare states valued has evolved into the economic system spectrum of capitalism to philanthropism, and the social concern that has been lacking in human economic history has been revealed. Thus, from SDGs, ESG, public goods, and the theory and practice related to public affairs of external effects, we can often find public governance mentioned by the Nobel Prize winner in Economics E. Ostrom. We can also see that the solution and tools of CSR are being applied to Taiwan's local development. As mentioned in this book: We have to learn how to become a B Corp, value SROI rather than only ROI, reinforce audit mechanism of SRI and common supervision by the people, and create human resource development mechanism based on SDG18. I believe such business management methodology let us pursue corporate development while ensuring the sustainability of the environment and human social survival.

Adjunct Assistant Professor of the Department of Finance of National Kaohsiung University of Science and Technology

Chairman of the Taiwan Industry-Academia-Research for Collaboration-Integration-Development Association

Professor of CSR Practical Courses in the Department of Finance of National Kaohsiung University of Science and Technology

PhD Degree in the Department of Public Affairs Management of National Sun Yat-sen University

Doctor Wei-Neng Huang

Reference

- Amy Luers (author). Yu-Hong Zheng from CSRone (translator). (2020, November 22). The Missing SDG: Ensure the Digital Age Supports People, Planet, Prosperity & Peace. https://csr.cw.com.tw/article/41743
- Tim Mohin, Chief Executive of GRI. Kai-Yuan Huang (translator). (2019, March 14). Tim Mohin: Corporate Sustainability Will Become Mainstream. https://csr.cw.com.tw/article/40894

A. International situation Analysis OECD-CSR

a. COVID-19 epidemic

COVID-19, the novel coronavirus epidemic is raging across the globe, and now over 50 countries around the world have detected a variant that's eager for action. The World Health Organization stated that given the virus's infection pattern, the epidemic in 2021 may be more severe than the first year. Its infectiousness will be stronger and still wide-spreading, setting off a huge tsunami-like wave on the epidemic prevention and economy of all countries. Looking back at the challenges of early 2020, they are like a trial to local governments. Of course, it impacts the entire world, and Taiwan's supply chain and economy are the most powerful that were never seen in history. Besides slowing the pace of global social and economic development, the virus also threatens the global system, including the loss of life, collapse to the financial system and the global crisis brought about by diversified political struggle. Apart from taking away precious lives and stagnating the economy, the "uncertainty" in political, social, and economic markets will become the norm in the future. It will surely shake the country to its core and seriously affect Taiwan's development if the small and medium-sized enterprises, which take on the heavy responsibility of Taiwan's industrial development, do not match for the outbreak of the epidemic and shut down. The outbreak has changed modern human values, but every coin has two sides. How will SMEs turn the crisis into an opportunity in the post-epidemic, economic recovery plan? The key to this book is the ability to quickly turn a crisis into an opportunity, adopt an appropriate layout in adversity, and mitigate the crisis with a business model; improving the scalability and flexibility of operations, providing the best way to increase not only economic but social benefits, looking ahead to the preemptive and business opportunities of the post-epidemic era are also the best times to address inequality. Even the entire world is now turbulent, enterprises in Taiwan are still very powerful and stable.

Beginning in the 1990s, a wave of globalization and trade liberalization hit the world, and now more than half of the world's top 10 economies are multinationals, with companies increasingly focusing on CSR from a strategic perspective, and influence has aligned with national forces. And the whole epidemic outbreak and enterprise risk control also need policy's guidance to pull to a higher level, CSR issue has become the driving force of a country's culture and brand. Statistics survey shows over 86% of enterprises believe that the practice of CSR is the ablest way to enhance the corporate. According to the research, intangible assets account for up to 80% of all business development value within an organization, showing the importance of corporate reputation to the development of its CSR operations. In view of this, many Taiwanese SMEs seek diversified business opportunities around the world, taking the global market as a battleground. Now CSR has formed a powerful force in the international community, the rise of national awareness has further affected the focus of government and corporate behavior trends. At this time, it is very important for enterprises and countries to increase mutual trust, strengthen their comprehensive and dynamic understanding and cooperation. Economic development statistics: when enterprises promote CSR affairs, the biggest dilemma is difficult to assess the benefits, enterprises do not know what to do or not to do. In particular, there is no obvious answer as to how the industry, the Government, and the academic community should be held accountable. However, the OECD multi-country corporate guidance program has a total of 10 guiding principles, the global issue under the Corporate Social Responsibility perspective, the true meaning of the macro CSR is in the moment of each implementation and action. TransAct is leading the digital trend to well use science and technology from optimization to transformation operation, the foresight to solve the strategic operation of government SMEs. By intelligent and efficient energy to bring more information to institutions and SMEs, through the global strategy of decision-making value to show the future of the new appearance.

b. Four shortcomings of enterprises

Global investment in CSR has already become significant learning. While facing the impact of the wave of epidemics, about 98% of SMEs in Taiwan can say to bear the brunt. Strengthening the competitiveness of the government and enterprises together is to ignite TransAct's innovation momentum. TransAct is committed to assisting and promises to provide business strategy momentum with the most professional service quality; let small, medium, and micro-enterprises that are willing to implement CSR keep up fast because CSR is not the patents of large companies. More SMEs in Taiwan need to take part in promotion "The proportion of small and medium-sized enterprises in all enterprises is 97.80%", the sudden and uncontrollable COVID-19 outbreak in 2020 let TransAct realized that SMEs are already in a hard time and many business operators believe that implementing CSR means that more enterprise resources must be invested and cause the operating costs to become higher, so many SMEs refused to invest in CSR during the promotion of CSR-related issues. TransAct understands that while enterprises are investing in CSR will encounter four scarcities, which is 1. Insufficient funding 2. Lack of knowledge 3. Not enough time 4. be short-handed. While supporting the SMEs, TransAct noticed that they aren't familiar with the government policies, they often misunderstand and take no count of the government's purpose, but in fact, the central and local governments are actively implementing those policies, and Taiwan's annual budgets for various ministries and conferences dedicated to supporting SMEs on production and operation management, marketing management, human resource management, research & development management, finance management, digital transformation and also New Southbound Policy. Because SMEs used to focus on the investment in their own industry and failed to understand the strategy side by side with national policies, they often face a new reality that is full of uncertainty and unpredictable.

c. Enterprise health check and Visa

The current business model of SMEs does not fully reflect the internalization of external costs, but who has a solution? TransAct does! TransAct communicates deeply with small, medium, and micro enterprises through its experience, and aims at the future challenges faced by SMEs, replacing "incremental change" with "systematic transformation", using systematic and more effective indicators to measure performance, and constructing the alliance works between SMEs and officials, academia, and research units to integrate gaps. A comprehensive solution with multiple interactions between various resources and various ministries: let the country and the company move towards international values and influence, maximize the benefits under the focus of efficiency, and explore the core of CSR with strength. Nowadays, nearly 80%-90% of foreign companies and listed companies choose suppliers based on their CSR performance. The intelligent guidance system integrated through the CSR business model of TransAct, so that companies can more easily condense business strategies and solve industrial human resources and talents, funds, R&D, technology, inheritance, resource matching, rewards and subsidies, industry information, internationalization related issues. The digital technology customization of TransAct provides practical application of enterprise innovative business models and implements specific strategies for business policy planning and CSR. Allow more SMEs to apply for national policies, central government plans, local government subsidies, and the transformational energy of digital technology. It also provides a full range of corporate CSR (ESGs) inspections and digital cloud information record retention, so that the Taiwan government and SMEs can benefit from the continuous development of its constitution; and through a third-party impartial agency: joint lawyers and accountants to do the annual CSR report endorsement visa, let a fulcrum drive the multiplied international economic benefits, actively create an innovative value chain industrial environment, and move towards globalization. The rapid development of technology and the market are far-reaching. The business model of TransAct will continue to plan short, medium, and long-term strategic goals towards the definition of corporate sustainable development, focusing on the widely accepted consensus definition and creating shared core values.

Reference

- Jia-Zheng Hua. (2021, January). The Prospect of Global Economy in 2021. http://www.cnfi.org.tw/front/bin/ptdetail.phtml?Part=magazine11001-610-2
- Ya-Hui Lin. (2006). The Situation of Corporate Social Responsibility Development in Taiwan (master's thesis). https://www.airitilibrary.com/Publication/alDetailed-Mesh1?DocID=U0006-0802200612125000
- Xiao-Wen Zhang (translator). Si-Qi Yan (sub-editor). (2021, January 14). New Variants of Coronavirus are More Contagious; WHO: The Pandemic Will be More Severe in the Second Year. https://www.cna.com.tw/news/aopl/202101140055.aspx
- BBC News. (2020, December 31). New Variants of Coronavirus Found in Asia; Countries Such as India, Korea, and Japan Place Restrictions on Immigration. https://www.bbc.com/zhongwen/trad/world-55496215

B. Company Establish.
TransAct, Inc.

┌───┐
【The Corporate Social Responsibilities index】

Goal 1: End poverty in all its forms everywhere.

Goal 2: End hunger, achieve food security and improved nutrition and promote sustainable agriculture.

Goal 3: Ensure healthy lives and promote well-being for all at all ages.

Goal 4: Ensure inclusive and equitable quality education and promote lifelong learning opportunities for all.

Goal 5: Achieve gender equality and empower all women and girls.

Goal 6: Ensure availability and sustainable management of water and sanitation for all.

Goal 7: Ensure access to affordable, reliable, sustainable and modern energy for all.

Goal 8: Promote sustained, inclusive and sustainable economic growth, full and productive employment and decent work for all.

Goal 9: Build resilient infrastructure, promote inclusive and sustainable industrialization, and foster innovation.

Goal 10: Reduce inequality within and among countries.

Goal 11: Make cities and human settlements inclusive, safe, resilient and sustainable.

Goal 12: Ensure sustainable consumption and production patterns.

Goal 13: Take urgent action to combat climate change and its impacts.

Goal 14: Conserve and sustainably use the oceans, seas and marine resources for sustainable development.

Goal 15: Protect, restore and promote sustainable use of terrestrial ecosystems, sustainably manage forests, combat desertification, and halt and reverse land degradation and halt biodiversity loss.

Goal 16: Promote peaceful and inclusive societies for sustainable development, provide access to justice for all and build effective, accountable and inclusive institutions at all levels.

Goal 17: Strengthen the means of implementation and revitalize the global partnership for sustainable development.

TransAct respond to international initiatives-

Goal 18: Ensuring the Digital Age Supports People and Planet, correct use of digital technology to promote peace and well-being of humankind.

a. Influence

(a) 【 Critical energy awaken by trends 】

TransAct will lead the country's SMEs to keep up with international trends, from never heard and don't understand about CSR across to understand CSR without experiencing the process that wants to but don't know how to do, because this is the era that enterprise makes strategic alliance and cooperation rather than being a single-player. TransAct strongly promote the CSR concept and use more innovative actions and strategic layout to drive industries, actively and concrete assessment and implementation methods and industrial models, and learning maps, so that SMEs and government can together attach the importance and carry out social responsibility.

(1) TransAct's digital technology platform will lead the whole SMEs in Taiwan and international markets from never heard and don't understand about CSR to execute CSR without experiencing the process that wants to but don't know how to do. Converged and provides all relevant international GRI SDGs, ESG, and B Corp certifications from domestic and abroad.

(2) Build up a knowledge base: let each enterprise find and demonstrate the uniqueness and specific core values of its own CSR in the knowledge base, our platform can connect to the network and search on the Internet for all relevant information.

(3) Facebook's teaching videos and step by step platform operations make intelligent learning models more indicator and advocate for the state to promote government agencies and businesses to become the most respected object of society.

(b) 【A key jump point in new economic model】

Through more than 25 years of experience, TransAct assists domestic SMEs and international enterprises to have deep communication, with the enterprise language to understand their real needs and corporate qualities to strengthen the implementation of the CSR program. To provide SMEs practical analysis and resource integration, promote the integration of production, officials, learning, research, and mutual efforts to share and create resources. TransAct provides SMEs "customized" innovative business models and applied practices so that when enterprises face the crisis of the epidemic can be transformed into diversity unlimited business opportunities, enterprise development blueprint can create a well-organized CSR integrated business strategy. State-to-local; local-to-national. This plan holds the key factor, allowing more SMEs to obtain policy drivers, central government plans, local government subsidies, industrial investment, key support, guidance energy, and momentum for digital technology transformation through national encouragement.

(1) Through the ESAA self-assessment form provided by TransAct, a more comprehensive comparison of domestic and foreign-related gaps and market development and subsidies will enable the organization to undergo a comprehensive and systematic transformation.

(2) SESP allows the management team and stakeholders of SMEs to quickly review and derive gaps and solutions that are most beneficial to business strategies.

(3) Provide the latest international trends, business opportunities, public benefits, and industry linkage opportunities to jointly solve corporate and social problems.

(c) 【The cloud digital passbook for the reserve of enterprise CSR】

To store the competitiveness after the epidemic with new business strategies and provide comprehensive corporate health check CSR (ESGs) and digital information for future reference. Let Taiwan government and SMEs have a good foundation for sustainable development and profit; TransAct focuses on deepening and combining a third-party impartial agency: joint lawyers and accountants to do the annual CSR report endorsement visa, let the government and SMEs CSR passbook gradually accumulate practical experience than by a fulcrum drive the multiplied international economic benefits, actively create an innovative value chain industrial environment, and finally illuminate the path of global development for SMEs.

(1) The cloud digital passbook can be downloaded at any time to provide the company's own latest CSR implementation and records, and it can also call out the evaluation data (discussions) of the previous period at any time as a reference for filling in the new period.

(2) Download and update at any time to provide the latest CSR tangible achievements and records planned by various departments of the enterprise.

b. Benefit

Provide enterprises with various resources and integrated applications to maintain enterprise's operations, promote joint research and development projects between five industries and academia, Social Affairs Bureau of Kaohsiung City Government website maintenance, Social Affairs Bureau of Kaohsiung City Government Children Welfare Service Center's website Construction, National taxation Bureau of Pingtung Branch policy disseminate, Kinmen Kaoliang Liquor Inc. staff retirement lecture, revitalize and increase the employed population, provide local employment opportunities, promote cross-industry alliances, promote entrepreneurship motivation of local population, regional digital cluster transformation counseling, respond and go to various companies to promote African shoe rescue operations, collect materials from various enterprises irregularly every month and send them to different units and organizations in various places, assist company employees to improve collaborative work, assist companies to promote employee functions and work efficiency, assist companies to promote employees' work-life balance, assist companies to improve the work environment and welfare of employees, assist employees to improve the comfort of office equipment and obtain happiness, assist companies to greatly reduce employee turnover, assist the company's employees to significantly improve their work efficiency, help companies increase the integrity between the company and employees, assist employees in working attitudes such as loyalty, assist enterprises to establish and unite employees' centripetal force, assist companies to establish brand LOGO and brand story marketing, assist in enhancing the image of the company, assist companies to develop international markets, assisting companies to set up new Southbound overseas companies, assist enterprises to deal with labor issues peacefully, assist companies to solve related legal issues, assist companies to improve its financial gaps, assist corporate tax clarification and adjustment, assisting enterprises in applying epidemic relief subsidies, assisting companies in applying epidemic relief loans, assist local youth in applying for youth entrepreneurship, assist enterprises in applying Micro-Business Startup Phoenix Program, Miss Xie Xin-hua, assisting companies to handle three major plans for investment in Taiwanese firms, assisting enterprises in applying SMEs Credit Guarantee Fund program, assist the development of the second generation succession transformation strategy of the enterprise.

c. TransAct Association NGO

The colleagues of TransAct worked together to establish the "Chinese Association of TransAct" in March 2020. Knowledge turns the future, TransAct book store gives free lessons to students under the poverty line, we are very gratified that students receive the President Educational Award in 2020. In 2020, tens of millions of donations be poured into the national education promotion knowledge because of the report by the weekly publication - *Business Today*. The educational quality of the Millennium Covenant and the concept of poverty eradication and equality have implemented the individual, organization, local community, and country.

The latest influence: Chinese Association of TransAct established in March 2020, assists TransAct second-hand bookstore to apply for art FUN coupons in response to the epidemic, and it successfully attracts tourists from all over the world, and successful fundraising student improvement award and scholarship fundraising. *Business Today* weekly magazine reads and forwards a lot in December 2020: insight into how we help those students who are under the poverty line and low academic achievement, provide them one-to-one free after-school tutoring; currently, each semester has an average of 170 students who's under the poverty line, from Minxiong, Chiayi City, Douliu...... and other regions. Chinese Association of TransAct poured in millions of donations from benevolent people across the country. Summarizing the above firmly believed concepts, TransAct continues to eradicate poverty in the region, eradicate hunger of school children, improve the scale of education quality, take care of children's health, and pay more attention to the drinking water and environmental sanitation in the after-school tutoring center. Knowledge lifts local employment and economic growth out of poverty and drives urban development towards the goal of endless economic learning.

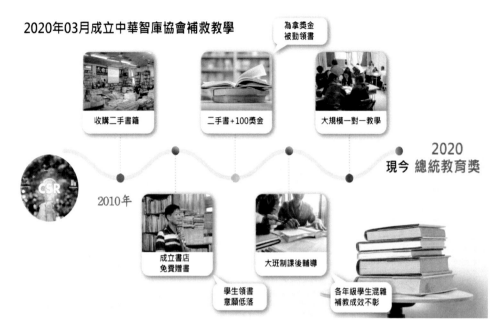

2020年03月成立中華智庫協會補救教學

為拿獎金
被動領書

收購二手書籍

二手書+100獎金

大規模一對一教學

2020
現今 總統教育獎

CSR

2010年

成立書店
免費贈書

大班制課後輔導

學生領書
意願低落

各年級學生混雜
補救成效不彰

March 2020: Chinese Association of TransAct remedial teaching established

2010: • Purchase second-hand books.

    • Set up a bookstore and give away books for free (But students have low willingness to receive books)

    • Second-hand books + 100 dollars' prize (Students got the book to get the bonus)

    • After-school tutoring in large class size (Results are ineffective because students are mixed with different grade)

    • After-school tutoring one in one

Now: 2020 President Educational Awards

Reference

- Jun-Tang Wang. (2020, December 10). Director Gave Up High Salary to Open a Book Store. https://www.businesstoday.com.tw/article/category/183035/post/202012100011/%E6%9C%80%E5%BC%B7%E9%A4%A8%E9%95%B7%E6%A3%84%E7%99%BE%E8%90%AC%E5%B9%B4%E8%96%AA%AA%EF%BC%81%20%E7%A0%B8%E5%8D%83%E8%90%AC%AC%E5%90%80%E4%BC%91%E9%87%91%E5%8F%AA%AA%E7%82%BA%E5%81%A0%BA%E4%E3%80%8D%EF%BC%81

- Grant Thornton. The Composition of Corporate Social Responsibility Report. https://www.grantthornton.tw/services/advisory-services/advisory-services2/

- Jie-Wen Li & Liang-Qu Wei. (2014, September). How to Enhance the Performance of "Corporate Social Responsibility". https://www2.deloitte.com/tw/tc/pages/risk/articles/newsletter-09-24.html

- Yu Shen. (2020, July 27). The CSR Spirit of CTBC Financial Holding: "Teaching a man how to fish is better than giving him a fish." https://www.gvm.com.tw/article/73856?utm_campaign=daily&utm_medium=social&utm_source=facebook&utm_content=GV_post&fbclid=IwAR3g9O_j4lPOGsRwPtyo7yLIK587SfCenI Ihp189Xqe44yQtmbEOqRS5oqNY

- iKnow. (2006, June 1). Statistics from White Paper On SMEsare Out. https://iknow.stpi.narl.org.tw/Post/Read.aspx?PostID=857

- Yi-Fan Gao. (2010, March 8). Does CSR Stay in the Hearts of Corporations and Citizens? https://www.gvm.com.tw/article/13760

- Yao-Hong Xu & Shi-Rong Zhan. (2006). Analysis of the Legal Nature of the OECD Guidelines for Multinational Enterprisesfrom the Perspective of International Law and the Adaptive Strategy of Taiwan. *Newsletter for Research of Applied Ethics*, 200611 (40 issues), 51-59. http://lawdata.com.tw/tw/detail.aspx?no=180445

- Industrial Sustainable Development Clearinghouse. (2016, March 7). About Corporate Social Responsibility. https://proj.ftis.org.tw/isdn2019/Application/Detail/6F4893445DDDBA15

C. SDGs Goal

a. GRI Standards become new standards for CSR worldwide

Global Reporting Initiative (GRI) issued a global standard for sustainability reporting in October 2016, providing organizations with a common language for public non-financial information. According to the regulations of the GRI organization, GRI Standards. GRI Standards will replace the G4 Guidelines from July 1, 2018, which can help organizations make better decisions and contribute to the value of the United Nations 2030 Sustainable Development Goals, and become a global CSR corporate society the new standard of responsibility reporting.

b. SDG Goal 18-Digital technology for the well-being of mankind and the planet

United Nations 18 Sustainable Development Goals (SDGs)

SDGs Goal 1   End poverty in all its forms everywhere.

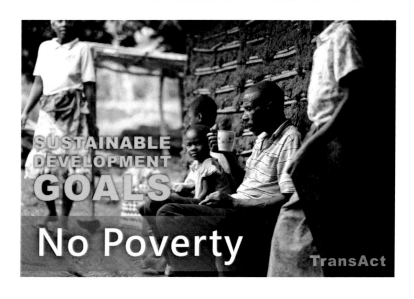

1.1 By 2030, eradicate extreme poverty for all people everywhere, currently measured as people living on less than $1.25 a day.

    1.1.1 Proportion of population below $1.25 (PPP) per day. [Disaggregated by sex, age, location (ex: city/countryside) and employment group.]

1.2 By 2030, reduce at least by half the proportion of men, women and children of all ages living in poverty in all its dimensions according to national definitions.

    1.2.1 Proportion of the population below the national poverty line. (Disaggregated by sex and age group.)

    1.2.2 Proportion of population living below national poverty line, disaggregated by sex and age group.

1.3 Implement nationally appropriate social protection systems and measures for all, including floors, and by 2030 achieve substantial coverage of the poor and the vulnerable.

    1.3.1 Percentage of population covered by social insurance programs. (Disaggregated by sex, with break down by children, unemployed, old age, people with disabilities, pregnant women/new-born, work injury victims, poor and vulnerable.)

1.4 By 2030, ensure that all men and women, in particular the poor and the vulnerable, have equal rights to economic resources, as well as access to basic services, ownership and control over land and other forms of property, inheritance, natural resources, appropriate new technology and financial services, including microfinance.

    1.4.1 Proportion of population living in the households with access to basic services.

    1.4.2 Proportion of adult population with tenure that is legally recognized. (Disaggregated by sex and type).

1.5 By 2030, build the resilience of the poor and those in vulnerable situations and reduce their exposure and vulnerability to climate-related extreme events and other economic, social and environmental shocks and disasters.

    1.5.1 Number of Deaths and Missing Persons Attributed to Disasters (per 100,000 people). (Same as 11.5.1).

1.5.2   Direct economic loss attributed to disasters in relation to global gross domestic product (GDP).

1.5.3   Number of countries with national and local disaster risk reduction strategies. (Same as 11.B.2)

1.A   Ensure significant mobilization of resources from a variety of sources, including through enhanced development cooperation, in order to provide adequate and predictable means for developing countries, in particular least developed countries, to implement programmes and policies to end poverty in all its dimensions.

1.A.1   Proportion of resources allocated by the government directly to poverty reduction programmes.

1.A.2   Proportion of total government spending on essential services (education, health and social protection).

1.B   Create sound policy frameworks at the national, regional and international levels, based on pro-poor and gender-sensitive development strategies, to support accelerated investment in poverty eradication actions.

1.B.1   Proportion of government recurrent and capital spending to sectors that disproportionately benefit women, the poor and vulnerable groups.

SDGs Goal 2   End hunger, achieve food security and improved nutrition and promote sustainable agriculture.

In 2015, the United Nations launched the Sustainable Development Goals 2030 (Sustainable Development Goals, SDGs) and proposed 17 core goals for global governments and enterprises to jointly move towards sustainable development and address issues such as the gap between rich and poor, climate change and gender parity. -- The second goal is "End hunger, achieve food security and improved nutrition and promote sustainable agriculture." What are the regulations and core spirit of this goal? Are there examples and reflections inside?

2.1 By 2030, end hunger and ensure access by all people, in particular the poor and people in vulnerable situations, including infants, to safe, nutritious and sufficient food all year round.

    2.1.1 Prevalence of undernourishment.

    2.1.2 Prevalence of moderate or severe food insecurity in the population, based on the food insecurity experience scale.

2.2 By 2030, end all forms of malnutrition, including achieving, by 2025, the internationally agreed targets on stunting and wasting in children under 5 years of age, and address the nutritional needs of adolescent girls, pregnant and lactating women and older persons.

    2.2.1 Prevalence of stunting among children under 5 years of age. (Plus, or minus 2 standard deviation from the median of the WHO Child Growth Standards.)

    2.2.2 Prevalence of malnutrition among children under 5 years of age, by type. (Disaggregated by wasting and overweight group.) (Plus, or minus 2 standard deviation from the median of the WHO Child Growth Standards.)

2.3 By 2030, double the agricultural productivity and incomes of small-scale food producers, in particular women, indigenous peoples, family farmers, pastoralists and fishers, including through secure and equal access to land, other productive resources and inputs, knowledge, financial services, markets and opportunities for value addition and non-farm employment.

    2.3.1 Volume of production per labor unit. (Disaggregated by classes of farming, pastoral and forestry enterprise size.)

2.3.2　Average income of small-scale food producers, by sex and indigenous status.

2.4　By 2030, ensure sustainable food production systems and implement resilient agricultural practices that increase productivity and production, that help maintain ecosystems, that strengthen capacity for adaptation to climate change, extreme weather, drought, flooding and other disasters and that progressively improve land and soil quality.

2.4.1　Proportion of agricultural area under productive and sustainable agriculture.

2.5　By 2020, maintain the genetic diversity of seeds, cultivated plants and farmed and domesticated animals and their related wild species, including through soundly managed and diversified seed and plant banks at the national, regional and international levels, and promote access to and fair and equitable sharing of benefits arising from the utilization of genetic resources and associated traditional knowledge, as internationally agreed.

2.5.1　Number of plant genetic resources for food and agriculture secured in medium- or long-term conservation facilities.

2.5.2　Proportion of local breeds classified as being at risk of extinction, not being at risk of extinction or unknown risk of extinction.

2.A　Increase investment, including through enhanced international cooperation, in rural infrastructure, agricultural research and extension services, technology development and plant and livestock gene banks in order to enhance agricultural productive capacity in developing countries, in particular least developed countries.

2.A.1　Agriculture orientation index for government expenditures.

2.A.2　Expenditure in the agricultural sector as a percentage of total government expenditure.

2.B　Correct and prevent trade restrictions and distortions in world agricultural markets, including through the parallel elimination of all forms of agricultural export subsidies and all export measures with equivalent effect, in accordance with the mandate of the Doha Development Round.

2.B.1　Evolution of potentially trade restrictive and distortive measures in agriculture.

2.B.2　Agricultural export subsidies.

2.C　Adopt measures to ensure the proper functioning of food commodity markets and their derivatives and facilitate timely access to market information, including on food reserves, in order to help limit extreme food price volatility.

2.C.1　Indicator of food price anomalies.

| SDGs Goal 3 | Ensure healthy lives and promote well-being for all at all ages. |

In 2015, the United Nations launched the Sustainable Development Goals 2030 (Sustainable Development Goals, SDGs) and proposed 17 core goals for global governments and enterprises to jointly move towards sustainable development and address issues such as the gap between rich and poor, climate change and gender parity. -- The third goal is "Ensure healthy lives and promote well-being for all at all ages." What are the regulations and core spirit of this goal? Are there examples and reflections inside?

3.1　By 2030, reduce the global maternal mortality ratio to less than 70 per 100,000 live births.

3.1.1　Maternal mortality ratio.

3.1.2　Proportion of births attended by skilled health personnel.

3.2 By 2030, end preventable deaths of newborns and children under 5 years of age, with all countries aiming to reduce neonatal mortality to at least as low as 12 per 1,000 live births and under-5 mortality to at least as low as 25 per 1,000 live births.

    3.2.1   Under-five mortality rate.

    3.2.2   Neonatal mortality rate.

3.3 By 2030, end the epidemics of AIDS, tuberculosis, malaria and neglected tropical diseases and combat hepatitis, water-borne diseases and other communicable diseases.

    3.3.1   Number of new HIV infections per 1,000 uninfected population. (Disaggregated by sex, age and key populations.)

    3.3.2   Tuberculosis incidence per 1,000 population.

    3.3.3   Malaria incidence per 1,000 population.

    3.3.4   Hepatitis B incidence per 100,000 population.

    3.3.5   Number of people requiring interventions against neglected tropical diseases.

3.4 By 2030, reduce by one third premature mortality from non-communicable diseases through prevention and treatment and promote mental health and well-being.

    3.4.1   Mortality rate attributed to cardiovascular disease, cancer, diabetes or chronic respiratory disease.

    3.4.2   Suicide mortality rate.

3.5 Strengthen the prevention and treatment of substance abuse, including narcotic drug abuse and harmful use of alcohol.

    3.5.1   Coverage of treatment interventions (pharmacological, psychosocial and rehabilitation and aftercare services) for substance use disorders.

    3.5.2   Harmful use of alcohol, defined according to the national context as alcohol per capita consumption (aged above 15 years and older) within a calendar year in liters of pure alcohol.

3.6 By 2020, half the number of global deaths and injuries from road traffic accidents.

    3.6.1   Death rate due to road traffic injuries.

3.7 By 2030, ensure universal access to sexual and reproductive health-care services, including for family planning, information and education, and the integration of reproductive health into national strategies and programmes.

    3.7.2   Adolescent birth rate per 1,000 women in that age group. (Aged 10-14 years, aged 15-19 years.)

3.8   Achieve universal health coverage, including financial risk protection, access to quality essential health-care services and access to safe, effective, quality and affordable essential medicines and vaccines for all.

    3.8.1   Coverage of essential health services. (Defined as the average coverage of essential services based on tracer interventions that include reproductive, maternal, newborn and child health, infectious diseases, non-communicable diseases and service capacity and access, among the general and the most disadvantaged population.)

    3.8.2   Proportion of population with large household expenditures on health as a share of total household expenditure or income.

3.9   By 2030, substantially reduce the number of deaths and illnesses from hazardous chemicals and air, water and soil pollution and contamination.

    3.9.1   Mortality rate attributed to household and ambient air pollution.

    3.9.2   Mortality rate attributed to unsafe water, unsafe sanitation and lack of hygiene.

    3.9.3   Mortality rate attributed to unintentional poisoning.

3.A   Strengthen the implementation of the World Health Organization Framework Convention on Tobacco Control in all countries, as appropriate.

    3.A.1   Age-standardized prevalence of current tobacco use among persons aged 15 years and older.

3.B   Support the research and development of vaccines and medicines for the communicable and non-communicable diseases that primarily affect developing countries, provide access to affordable essential medicines and vaccines, in accordance with the Doha Declaration on the TRIPS Agreement and public health, which affirms the right of developing countries to use to the full the provisions in the agreement on Trade-related Aspects of Intellectual Property Rights regarding flexibilities to protect public health, and, in particular, provide access to medicines for all.

    3.B.1   Proportion of the population with access to affordable medicines and vaccines on a sustainable basis.

3.B.2   Total net official development assistance to medical research and basic health sectors.

3.C   Substantially increase health financing and the recruitment, development, training and retention of the health workforce in developing countries, especially in least developed countries (LIDs) and small island developing states (SIDs).

3.C.1   Health worker density and distribution.

3.D   Strengthen the capacity of all countries, in particular developing countries, for early warning, risk reduction and management of national and global health risks.

3.D.1   International Health Regulations (IHR) capacity and health emergency preparedness.

SDGs Goal 4   Ensure inclusive and equitable quality education and promote lifelong learning opportunities for all.

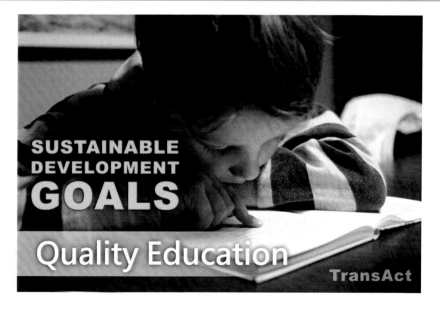

In 2015, the United Nations launched the Sustainable Development Goals 2030 (Sustainable Development Goals, SDGs) and proposed 17 core goals for global governments and enterprises to jointly move towards sustainable development and address issues such as the gap between rich and poor, climate change and gender parity. --

The third goal is "Ensure inclusive and equitable quality education and promote lifelong learning opportunities for all." What are the regulations and core spirit of this goal? Are there examples and reflections inside?

4.1 By 2030, ensure that all girls and boys complete free, equitable and quality primary and secondary education leading to relevant and effective learning outcomes.

    4.1.1 Proportion of children and young people achieving at least a minimum proficiency level in reading and mathematics. (Disaggregated by sex, and (a) in grades 2/3; (b) at the end of primary; and (c) at the end of lower secondary.)

4.2 By 2030, ensure that all girls and boys have access to quality early childhood development, care and pre-primary education so that they are ready for primary education.

    4.2.1 Proportion of children under 5 years of age who are developmentally on track in health, learning and psychosocial well-being. (Disaggregated by sex.)

    4.2.2 Participation rate in organized learning (In terms of one year before the official primary entry age and disaggregated by sex.)

4.3 By 2030, ensure equal access for all women and men to affordable and quality technical, vocational and tertiary education, including university.

    4.3.1 Participation rate of youth and adults in formal and non-formal education and training in the previous 12 months. (Disaggregated by sex.)

4.4 By 2030, substantially increase the number of youth and adults who have relevant skills, including technical and vocational skills, for employment, decent jobs and entrepreneurship.

    4.4.1 Proportion of youth and adults with information and communications technology (ICT) skills, by type of skill.

4.5 By 2030, eliminate gender disparities in education and ensure equal access to all levels of education and vocational training for the vulnerable, including persons with disabilities, indigenous peoples and children in vulnerable situations.

    4.5.1 Parity indices for all education indicators. (Disaggregated by female/

male, rural/urban, bottom/top wealth quintile and others such as disability status, indigenous peoples and conflict-affected, as data become available.)

4.6 By 2030, ensure that all youth and a substantial proportion of adults, both men and women, achieve literacy and numeracy.

4.6.1 Percentage of population in a given age group achieving at least a fixed level of proficiency in functional literacy and numeracy skills. (Disaggregated by sex.)

4.7 By 2030, ensure that all learners acquire the knowledge and skills needed to promote sustainable development, including, among others, through education for sustainable development and sustainable lifestyles, human rights, gender equality, promotion of a culture of peace and non-violence, global citizenship and appreciation of cultural diversity and of culture's contribution to sustainable development.

4.7.1 Extent to which global citizenship education and education for sustainable development, including gender equality and human rights, are mainstreamed at all levels in: (a) national education policies, (b) curricula, (c) teacher education and (d) student assessment.

4.A Build and upgrade education facilities that are child, disability and gender sensitive and provide safe, non-violent, inclusive and effective learning environments for all.

4.A.1 Proportion of schools with access to: (a) electricity; (b) the Internet for pedagogical purposes; (c) computers for pedagogical purposes; (d) adapted infrastructure and materials for students with disabilities; (e) basic drinking water; (f) single-sex basic sanitation facilities; and (g) basic handwashing facilities. (Base on the WASH indicator definitions.)

4.B By 2020, substantially expand globally the number of scholarships available to developing countries, in particular least developed countries, small island developing States and African countries, for enrolment in higher education, including vocational training and information and communications technology, technical, engineering and scientific programmes, in developed countries and other developing countries.

    4.B.1   Volume of official development assistance flows for scholarships. (Disaggregated by sector and type of study.)

4.C   By 2030, substantially increase the supply of qualified teachers, including through international cooperation for teacher training in developing countries, especially least developed countries and small island developing States.

    4.C.1   Proportion of teachers in: (a) pre-primary; (b) primary; (c) lower secondary; and (d) upper secondary education who have received at least the minimum organized teacher training (e.g., pedagogical training) pre-service or in-service required for teaching at the relevant level in a given country.

SDGs Goal 5   Achieve gender equality and empower all women and girls.

In 2015, the United Nations launched the Sustainable Development Goals 2030 (Sustainable Development Goals, SDGs) and proposed 17 core goals for global governments and enterprises to jointly move towards sustainable development and address issues such as the gap between rich and poor, climate change and gender parity. -- The third goal is "Achieve gender equality and empower all women and girls." What are the regulations and core spirit of this goal? Are there examples and reflections inside?

5.1   End all forms of discrimination against all women and girls everywhere.

    5.1.1   Whether or not legal frameworks are in place to promote, enforce and monitor equality and non-discrimination on the basis of sex.

5.2   Eliminate all forms of violence against all women and girls in the public and private spheres, including trafficking and sexual and other types of exploitation.

    5.2.1   Proportion of ever-partnered women and girls aged 15 years and older subjected to physical, sexual or psychological violence by a current or former intimate partner in the previous 12 months, by form of violence and by age.

    5.2.2   Proportion of women and girls aged 15 years and older subjected to sexual violence by persons other than an intimate partner in the previous 12 months, by age and place of occurrence.

5.3   Eliminate all harmful practices, such as child, early and forced marriage and female genital mutilation.

    5.3.1   Proportion of women aged 20-24 years who were married or in a union before age 15 and before age 18.

5.4   Recognize and value unpaid care and domestic work through the provision of public services, infrastructure and social protection policies and the promotion of shared responsibility within the household and the family as nationally appropriate.

    5.4.1   Proportion of time spent on unpaid domestic and care work, by sex, age and location.

5.5   Ensure women's full and effective participation and equal opportunities for leadership at all levels of decision-making in political, economic and public life.

    5.5.1   Proportion of seats held by women in national parliaments and local governments.

    5.5.2   Proportion of women in managerial positions.

5.6   Ensure universal access to sexual and reproductive health and reproductive rights as agreed in accordance with the Programme of Action of the International

Conference on Population and Development and the Beijing Platform for Action and the outcome documents of their review conferences.

5.6.1   Proportion of women aged 15-49 years who make their own informed decisions regarding sexual relations, contraceptive use and reproductive health care.

5.6.2   Number of countries with laws and regulations that guarantee women aged 15-49 years access to sexual and reproductive health care, information and education.

5.A   Undertake reforms to give women equal rights to economic resources, as well as access to ownership and control over land and other forms of property, financial services, inheritance and natural resources, in accordance with national laws.

5.A.1   (a) Percentage of people with ownership or secure rights over agricultural land (out of total agricultural population). (Disaggregated by sex.); and (b) share of women among owners or rights-bearers of agricultural land. (Disaggregated by type of tenure.)

5.A.2   Proportion of countries where the legal framework guarantees women's equal rights to land ownership or control.

5.B   Enhance the use of enabling technology, in particular information and communications technology, to promote the empowerment of women.

5.B.1   Proportion of individuals who own a mobile telephone. (Disaggregated by sex.)

5.C   Adopt and strengthen sound policies and enforceable legislation for the promotion of gender equality and the empowerment of all women and girls at all levels.

5.C.1   Proportion of countries with systems to track and make public allocations for gender equality and women's empowerment.

| SDGs Goal 6 | Ensure availability and sustainable management of water and sanitation for all. |

SUSTAINABLE DEVELOPMENT GOALS

Clean Water and Sanitation

TransAct

In 2015, the United Nations launched the Sustainable Development Goals 2030 (Sustainable Development Goals, SDGs) and proposed 17 core goals for global governments and enterprises to jointly move towards sustainable development and address issues such as the gap between rich and poor, climate change and gender parity. -- The third goal is "Ensure availability and sustainable management of water and sanitation for all." What are the regulations and core spirit of this goal? Are there examples and reflections inside?

6.1　By 2030, achieve universal and equitable access to safe and affordable drinking water for all.

　　6.1.1　Proportion of population using safely managed drinking water services.

6.2　By 2030, achieve access to adequate and equitable sanitation and hygiene for all and end open defecation, paying special attention to the needs of women and girls and those in vulnerable situations.

　　6.2.1　Proportion of population using safely managed sanitation services, including a hand-washing facility with soap and water.

6.3　By 2030, improve water quality by reducing pollution, eliminating dumping and minimizing release of hazardous chemicals and materials, halving the proportion of untreated wastewater and substantially increasing recycling and safe reuse globally.

6.3.1 Proportion of wastewater safely treated.

6.3.2 Proportion of bodies of water with good ambient water quality.

6.4 By 2030, substantially increase water-use efficiency across all sectors and ensure sustainable withdrawals and supply of freshwater to address water scarcity and substantially reduce the number of people suffering from water scarcity.

6.4.1 Change in water use efficiency over time.

6.4.2 Level of water stress: freshwater withdrawal as a proportion of available freshwater resources.

6.5 By 2030, implement integrated water resources management at all levels, including through transboundary cooperation as appropriate.

6.5.1 Degree of integrated water resources management implementation. (0-100)

6.5.2 Proportion of transboundary basin area with an operational arrangement for water cooperation.

6.6 By 2020, protect and restore water-related ecosystems, including mountains, forests, wetlands, rivers, aquifers and lakes.

6.6.1 Change in the extent of water-related ecosystems over time.

6.A By 2030, expand international cooperation and capacity-building support to developing countries in water- and sanitation-related activities and programmes, including water harvesting, desalination, water efficiency, wastewater treatment, recycling and reuse technologies.

6.A.1 Amount of water- and sanitation-related official development assistance that is part of a government-coordinated spending plan.

6.B Support and strengthen the participation of local communities in improving water and sanitation management.

6.B.1 Proportion of local administrative units with established and operational policies and procedures for participation of local communities in water and sanitation management.

SDGs Goal 7 | Ensure access to affordable, reliable, sustainable and modern energy for all.

In 2015, the United Nations launched the Sustainable Development Goals 2030 (Sustainable Development Goals, SDGs) and proposed 17 core goals for global governments and enterprises to jointly move towards sustainable development and address issues such as the gap between rich and poor, climate change and gender parity. -- The third goal is "Ensure access to affordable, reliable, sustainable and modern energy for all." What are the regulations and core spirit of this goal? Are there examples and reflections inside?

7.1 By 2030, ensure universal access to affordable, reliable and modern energy services.

   7.1.1 Proportion of population with access to electricity.

   7.1.2 Proportion of population with primary reliance on clean fuels and technology.

7.2 By 2030, increase substantially the share of renewable energy in the global energy mix.

   7.2.1 Renewable energy share in the total final energy consumption.

7.3   By 2030, double the global rate of improvement in energy efficiency.

    7.3.1   Energy intensity measured in terms of primary energy and GDP.

7.A   By 2030, enhance international cooperation to facilitate access to clean energy research and technology, including renewable energy, energy efficiency and advanced and cleaner fossil-fuel technology, and promote investment in energy infrastructure and clean energy technology.

    7.A.1   International financial flows to developing countries in support of clean energy research and development and renewable energy production, including in hybrid systems.

7.B   By 2030, expand infrastructure and upgrade technology for supplying modern and sustainable energy services for all in developing countries, in particular least developed countries, small island developing States, and land-locked developing countries, in accordance with their respective programmes of support.

    7.B.1   Investments in energy efficiency as a percentage of GDP and the amount of foreign direct investment in financial transfer for infrastructure and technology to sustainable development services.

SDGs Goal 8   Promote sustained, inclusive and sustainable economic growth, full and productive employment and decent work for all.

In 2015, the United Nations launched the Sustainable Development Goals 2030 (Sustainable Development Goals, SDGs) and proposed 17 core goals for global governments and enterprises to jointly move towards sustainable development and address issues such as the gap between rich and poor, climate change and gender parity. -- The third goal is "Promote sustained, inclusive and sustainable economic growth, full and productive employment and decent work for all." What are the regulations and core spirit of this goal? Are there examples and reflections inside?

8.1   Sustain per capita economic growth in accordance with national circumstances and, in particular, at least 7% gross domestic product growth per annum in the least developed countries.

    8.1.1   Annual growth rate of real GDP per capita.

8.2   Achieve higher levels of economic productivity through diversification, technological upgrading and innovation, including through a focus on high-value added and labor-intensive sectors.

    8.2.1   Annual growth rate of real GDP per employed person.

8.3   Promote development-oriented policies that support productive activities, decent job creation, entrepreneurship, creativity and innovation, and encourage the formalization and growth of micro-, small- and medium-sized enterprises, including through access to financial services.

    8.3.1   Proportion of informal employment in non-agriculture employment. (Disaggregated by sex.)

8.4   Improve progressively, through 2030, global resource efficiency in consumption and production and endeavor to decouple economic growth from environmental degradation, in accordance with the 10-year framework of programmes on sustainable consumption and production, with developed countries taking the lead.

    8.4.1   Material footprint, material footprint per capita, and material footprint per GDP.

    8.4.2   Domestic material consumption, domestic material consumption per capita, and domestic material consumption per GDP.

8.5   By 2030, achieve full and productive employment and decent work for all women and men, including for young people and persons with disabilities, and equal pay for work of equal value.

8.5.1　Average hourly earnings of female and male employees. (Disaggregated by occupation, age and persons with disabilities.)

8.5.2　Unemployment rate. (Disaggregated by sex, age and persons with disabilities.)

8.6　By 2020, substantially reduce the proportion of youth not in employment, education or training.

8.6.1　Proportion of youth not in education, employment or training. (Aged between 15 and 24 years.)

8.7　Take immediate and effective measures to eradicate forced labor, end modern slavery and human trafficking and secure the prohibition and elimination of the worst forms of child labor, including recruitment and use of child soldiers, and by 2025 end child labor in all its forms.

8.7.1　Proportion and number of children aged between 5 to 17 years engaged in child labor. (Disaggregated by sex and age.)

8.8　Protect labor rights and promote safe and secure working environments for all workers, including migrant workers, in particular women migrants, and those in precarious employment.

8.8.1　Frequency rates of fatal and non-fatal occupational injuries. (Disaggregated by sex and migrant status.)

8.8.2　Increase in national compliance of labor rights (freedom of association and collective bargaining) based on International Labor Organization (ILO) textual sources and national legislation. (Disaggregated by sex and migrant status.)

8.9　By 2030, devise and implement policies to promote sustainable tourism that creates jobs and promotes local culture and products.

8.9.1　Tourism direct GDP as a proportion of total GDP and in growth rate.

8.9.2　Number of jobs in tourism industries as a proportion of total jobs and growth rate of jobs. (Disaggregated by sex.)

8.10　Strengthen the capacity of domestic financial institutions to encourage and expand access to banking, insurance and financial services for all.

8.10.1 Number of commercial bank branches and automated teller machines (ATMs) per 100,000 adults.

8.10.2 Proportion of adults (15 years and older) with an account at a bank or other financial institution or with a mobile-money-service provider.

8.A Increase Aid for Trade support for developing countries, in particular least developed countries, including through the Enhanced Integrated Framework for Trade-Related Technical Assistance to Least Developed Countries.

8.A.1 Aid for Trade commitments and disbursements.

8.B By 2020, develop and operationalize a global strategy for youth employment and implement the Global Jobs Pact of the International Labor Organization.

8.B.1 Total government spending in social protection and employment programmes as a proportion of the national budgets and GDP.

| SDGs Goal 9 | Build resilient infrastructure, promote inclusive and sustainable industrialization and foster innovation. |

In 2015, the United Nations launched the Sustainable Development Goals 2030 (Sustainable Development Goals, SDGs) and proposed 17 core goals for global governments and enterprises to jointly move towards sustainable development and address issues such as the gap between rich and poor, climate change and gender parity. -- The third goal is "Build resilient infrastructure, promote inclusive and sustainable industrialization and foster innovation." What are the regulations and core spirit of this goal? Are there examples and reflections inside?

9.1   Develop quality, reliable, sustainable and resilient infrastructure, including regional and transborder infrastructure, to support economic development and human well-being, with a focus on affordable and equitable access for all.

    9.1.1   Proportion of the rural population who live within 2 km of an all-season road.

    9.1.2   Passenger and freight volumes, by mode of transport.

9.2   Promote inclusive and sustainable industrialization and, by 2030, significantly raise industry's share of employment and gross domestic product, in line with national circumstances, and double its share in least developed countries.

    9.2.1   Manufacturing value added as a proportion of GDP and per capita.

    9.2.2   Manufacturing employment as a proportion of total employment.

9.3   Increase the access of small-scale industrial and other enterprises, in particular in developing countries, to financial services, including affordable credit, and their integration into value chains and markets.

    9.3.1   Proportion of small-scale industries in total industry value added.

    9.3.2   Proportion of small-scale industries with a loan or line of credit.

9.4   By 2030, upgrade infrastructure and retrofit industries to make them sustainable, with increased resource-use efficiency and greater adoption of clean and environmentally sound technologies and industrial processes, with all countries taking action in accordance with their respective capabilities.

    9.4.1 CO_2 emission per unit of value added.

9.5   Enhance scientific research, upgrade the technological capabilities of industrial sectors in all countries, in particular developing countries, including, by 2030, encouraging innovation and substantially increasing the number of research and development workers per 1 million people and public and private research and development spending.

9.5.1 Research and development expenditure as a proportion of GDP.

9.5.2 Researchers in full-time equivalent per million inhabitants.

9.A Facilitate sustainable and resilient infrastructure development in developing countries through enhanced financial, technological and technical support to African countries, least developed countries, landlocked developing countries and small island developing States.

9.A.1 Total official international support (official development assistance plus other official flows) to infrastructure.

9.B Support domestic technology development, research and innovation in developing countries, including by ensuring a conducive policy environment for, inter alia, industrial diversification and value addition to commodities.

9.B.1 Proportion of medium and high-tech industry value added in total value added.

9.C Significantly increase access to information and communications technology and strive to provide universal and affordable access to the Internet in least developed countries by 2020.

9.C.1 Proportion of population covered by mobile network.

SDGs Goal 10 Reduce inequality within and among countries.

In 2015, the United Nations launched the Sustainable Development Goals 2030 (Sustainable Development Goals, SDGs) and proposed 17 core goals for global governments and enterprises to jointly move towards sustainable development and address issues such as the gap between rich and poor, climate change and gender parity. -- The third goal is "Reduce inequality within and among countries." What are the regulations and core spirit of this goal? Are there examples and reflections inside?

10.1　By 2030, progressively achieve and sustain income growth of the bottom 40 per cent of the population at a rate higher than the national average.

　　10.1.1　Growth rates of household expenditure or income per capita among the bottom 40% of the population and the total population.

10.2　By 2030, empower and promote the social, economic and political inclusion of all, irrespective of age, sex, disability, race, ethnicity, origin, religion or economic or other status.

　　10.2.1　Proportion of people living below 50% of median income. (Disaggregated by age, sex and persons with disabilities.)

10.3　Ensure equal opportunity and reduce inequalities of outcome, including by eliminating discriminatory laws, policies and practices and promoting appropriate legislation, policies and action in this regard.

　　10.3.1　Proportion of the population reporting having personally felt discriminated against or harassed within the previous 12 months on the basis of a ground of discrimination prohibited under international human rights law. (Same as 16.B.1)

10.4　Adopt policies, especially fiscal, wage and social protection policies, and progressively achieve greater equality.

　　10.4.1　Labor share of GDP, comprising wages and social protection transfers.

10.5　Improve the regulation and monitoring of global financial markets and institutions and strengthen the implementation of such regulations.

　　10.5.1　Financial Soundness Indicators.

10.6　Ensure enhanced representation and voice for developing countries in decision-making in global international economic and financial institutions in order to deliver more effective, credible, accountable and legitimate institutions.

   10.6.1   Proportion of members and voting rights of developing countries in international organizations. (Same as 16.8.1)

10.7   Facilitate orderly, safe, regular and responsible migration and mobility of people, including through the implementation of planned and well-managed migration policies.

   10.7.1   Recruitment cost borne by employee as a proportion of yearly income earned in country of destination.

   10.7.2   Number of countries that have implemented well-managed migration policies.

10.A   Implement the principle of special and differential treatment for developing countries, in particular least developed countries, in accordance with World Trade Organization agreements.

   10.A.1   Proportion of tariff lines applied to imports from least developed countries and developing countries with zero-tariff.

10.B   Encourage official development assistance and financial flows, including foreign direct investment, to States where the need is greatest, in particular least developed countries, African countries, small island developing States and landlocked developing countries, in accordance with their national plans and programmes.

   10.B.1   Total resource flows for development, by recipient and donor countries and type of flow. (e.g., official development assistance, foreign direct investment and other flows)

10.C   By 2030, reduce to less than 3% the transaction costs of migrant remittances and eliminate remittance corridors with costs higher than 5%.

   10.C.1   Remittance costs as a proportion of the amount remitted.

SDGs Goal 11	Develop urban and rural areas that show characteristics of fusion, safety, resilience, and sustainability.

In 2015, the United Nations launched the Sustainable Development Goals 2030 (Sustainable Development Goals, SDGs) and proposed 17 core goals for global governments and enterprises to jointly move towards sustainable development and address issues such as the gap between rich and poor, climate change and gender parity. -- The third goal is "Develop urban and rural areas that show characteristics of fusion, safety, resilience, and sustainability." What are the regulations and core spirit of this goal? Are there examples and reflections inside?

11.1　By 2030, ensure access for all to adequate, safe and affordable housing and basic services and upgrade slums.

　　11.1.1　Proportion of urban population living in slums, informal settlements or inadequate housing.

11.2　By 2030, provide access to safe, affordable, accessible and sustainable transport systems for all, improving road safety, notably by expanding public transport, with special attention to the needs of those in vulnerable situations, women, children, persons with disabilities and older persons.

　　11.2.1　Proportion of population that has convenient access to public transport.

(Disaggregated by sex, age and persons with disabilities.)

11.3 By 2030, enhance inclusive and sustainable urbanization and capacity for participatory, integrated and sustainable human settlement planning and management in all countries.

    11.3.1 Ratio of land consumption rate to population growth rate.

    11.3.2 Proportion of cities with a direct participation structure of civil society in urban planning and management that operate regularly and democratically.

11.4 Strengthen efforts to protect and safeguard the world's cultural and natural heritage.

    11.4.1 Total expenditure (public and private) per capita spent on the preservation, protection and conservation of all cultural and natural heritage. [Disaggregated by type of heritage (cultural, natural, mixed and World Heritage Centre designation), level of government (national, regional and local/municipal), type of expenditure (operating expenditure/investment) and type of private funding (donations in kind, private non-profit sector and sponsorship.)]

11.5 By 2030, significantly reduce the number of deaths and the number of people affected and substantially decrease the direct economic losses relative to global gross domestic product caused by disasters, including water-related disasters, with a focus on protecting the poor and people in vulnerable situations.

    11.5.1 Number of deaths, missing persons and persons affected by disaster per 100,000 people. (Same as 1.5.1, 13.1.2)

    11.5.2 Direct disaster economic loss in relation to global GDP, including disaster damage to critical infrastructure and disruption of basic service.

11.6 By 2030, reduce the adverse per capita environmental impact of cities, including by paying special attention to air quality and municipal and other waste management.

    11.6.1 Proportion of urban solid waste regularly collected and with adequate final discharge out of total urban solid waste generated.

    11.6.2 Annual mean levels of fine particulate matter (e.g., $PM_{2.5}$ and PM_{10}) in cities. (population weighted)

11.7   By 2030, provide universal access to safe, inclusive and accessible, green and public spaces, in particular for women and children, older persons and persons with disabilities.

    11.7.1   Average share of the built-up area of cities that is open space for public use for all. (Disaggregated by sex, age and persons with disabilities.)

    11.7.2   Proportion of persons victim of physical or sexual harassment, in the previous 12 months. (Disaggregated by sex, age and persons with disabilities.)

11.A   Support positive economic, social and environmental links between urban, per-urban and rural areas by strengthening national and regional development planning.

    11.A.1   Proportion of population living in cities that implement urban and regional development plans integrating population projections and resource needs. (Disaggregated by size of city.)

11.B   By 2020, substantially increase the number of cities and human settlements adopting and implementing integrated policies and plans towards inclusion, resource efficiency, mitigation and adaptation to climate change, resilience to disasters, and develop and implement, according to the Sendai Framework for Disaster Risk Reduction 2015-2030, holistic disaster risk management at all levels.

    11.B.1   Proportion of local governments that adopt and implement local disaster risk reduction strategies according to the Sendai Framework for Disaster Risk Reduction 2015-2030.

    11.B.2   Number of countries with national and local disaster risk reduction strategies. (Same as 1.5.3)

11.C   Support least developed countries(LDCs), including through financial and technical assistance, in building sustainable and resilient buildings utilizing local materials.

    11.C.1   Proportion of national financial support to the least developed countries. (Using retrofitting of sustainable, resilient and resource-efficient buildings utilizing local material.)

SDGs Goal 12 Ensure sustainable consumption and production patterns.

In 2015, the United Nations launched the Sustainable Development Goals 2030 (Sustainable Development Goals, SDGs) and proposed 17 core goals for global governments and enterprises to jointly move towards sustainable development and address issues such as the gap between rich and poor, climate change and gender parity. -- The third goal is "Ensure sustainable consumption and production patterns." What are the regulations and core spirit of this goal? Are there examples and reflections inside?

12.1 Implement the green factory system, promote the Cradle to Cradle (C2C) design concept, encourage enterprises to produce green and low-carbon products, establish green standards for products and cleaner production, and actively implement the relocation of polluting factories to industrial parks.

    12.1.1 The number of companies that have passed the clean production compliance assessment of green factories.

    12.1.2 Draft the number of cradle-to-cradle design guidelines for industry promotion.

12.2 Master the use of key materials, incorporate them into the sustainable management of the material life cycle, and promote the sustainable use of raw materials.

       12.2.1   The distribution of key materials and the number of energy resources used.

       12.2.2   Resource productivity.

       12.2.3   Material consumption per capita.

12.3   Reduce food loss in the production supply chain and reduce food waste at the consumer end.

       12.3.1   Global Food Loss and Waste. (Vegetable/Fruit)

       12.3.2   Rate of food processing loss.

       12.3.3   The amount of food waste produced in supermarkets and retail stores.

12.4   Reduce waste through green production, improve waste recycling and treatment technology capabilities, promote the development of the resource recycling industry towards higher efficiency, and manage chemical substances and waste in accordance with international conventions.

       12.4.1   Proper reuse of enterprise waste.

       12.4.2   Recycling rate of industrial waste and output value of resource recycling industry.

       12.4.3   Recycling rate of industrial waste in science parks.

       12.4.4   Number of chemical substance flow tracking cases.

       12.4.5   Refer to the Stockholm Convention to handle the announcement of the number of poison inspections.

       12.4.6   The amount of hazardous industrial waste per capita.

       12.4.7   Number of industrial wastes incinerated and buried.

12.5   Promote the cross-industry cooperation chain model, integrate energy resources for effective recycling, and promote the development of national circular economy.

       12.5.1   The recycling rate of regional energy resources.

       12.5.2   Circular economy-industrial innovative chemical materials.

12.6   Encourage companies to adopt sustainable development measures and disclose sustainable development information, while ensuring the accuracy and quality of the information.

12.6.1   The number of product carbon footprint label certificates issued.

12.6.2   The amount of domestic bank financing for green industries.

12.6.3   The number of listed and OTC companies that compulsorily compile corporate social responsibility reports.

12.7   Promote public and private sectors to increase green procurement.

12.7.1   Green procurement ratio of government agencies.

12.7.2   The amount of green procurement by private enterprise.

12.8   Promote environmentally friendly and circular agriculture to reduce the pollution of soil and water caused by agricultural practices and waste generated.

12.8.1   Area that promote environmentally friendly and organic farming.

12.8.2   Recycling of livestock waste.

12.8.3   Recycling of fishery waste.

12.A   Support developing countries to strengthen their scientific and technological capacity to move towards more sustainable patterns of consumption and production.

12.A.1   Amount of support to developing countries on research and development for sustainable consumption and production and environmentally sound technologies.

12.B   Promote sustainable tourism development, guide the tourism industry to provide green and local tourism models, create a sustainable tourism environment in Taiwan and enhance the value of the industry.

12.B.1   The overall income growth rate of tourism.

12.B.2   The growth rate of employment in the tourism industry.

12.B.3   Draw up green tourism standards.

> **SDGs Goal 13** Take urgent action to combat climate change and its impacts.

In 2015, the United Nations launched the Sustainable Development Goals 2030 (Sustainable Development Goals, SDGs) and proposed 17 core goals for global governments and enterprises to jointly move towards sustainable development and address issues such as the gap between rich and poor, climate change and gender parity. -- The third goal is "Take urgent action to combat climate change and its impacts." What are the regulations and core spirit of this goal? Are there examples and reflections inside?

13.1   Improve the ability to adapt to climate change, strengthen resilience and reduce vulnerability.

    13.1.1   Make an inventory of climate risks and formulate priority action plans for adaptation and implementation accordingly.

13.2   Implement the greenhouse gas phase control targets.

    13.2.1   Achieve the control targets for each phase of greenhouse gas.

13.3   Improve climate change sustainable education and public literacy.

    13.3.1   Promote education on climate change and sustainable campuses.

    13.3.2   Promote changes in national behavior and implement low-carbon local actions.

13.3.3 Adapt to the construction and service of scientific capabilities in response to climate change.

SDGs Goal 14	Conserve and sustainably use the oceans, seas and marine resources, and prevent the ocean environment from deteriorating.

14.1 Reduce all kinds of marine pollution, including nutrients and marine debris.

    14.1.1 The eutrophication index and the amount of drifting plastic in the coastal area.

    14.1.2 The qualification rate of 7 water quality items including dissolved oxygen, heavy metal cadmium, lead, mercury, copper, zinc, and ammonia nitrogen in the national marine environmental water quality monitoring station.

14.2 Manage and protect marine and coastal ecology in a sustainable manner.

    14.2.1 The number of ocean areas where the concept of ecosystem management is used for resource management.

    14.2.2 Mean Nutrition Level (MTL) and Fishing-in-balance index (FiB).

    14.2.3 Establish a marine database.

14.3 Slow down and improve the impact of ocean acidification.

14.3.1　The average ocean pH value of the approved sampling site.

14.4　Effectively supervise harvesting, eliminate overfishing, illegal, unreported and unregulated (IUU), or destructive fishing practices, and try to restore fish stocks to sustainable development levels.

14.4.1　Resource management of economic fish species along coastal.

14.4.2　Effectively supervise harvesting, eliminate overfishing, and illegal, unreported and unregulated (IUU) fishing practices.

14.4.3　Subsidies to fishing boat operators to install Vessel Monitoring System (VMS) and other position report equipment to prevent blocking the proportion of illegal, unreported and unregulated fishing activities.

14.5　Protect at least 10% of coastal and marine areas.

14.5.1　The area of marine reserve areas accounts for the proportion of marine areas in our country.

14.5.2　The area of coastal protection areas accounts for the proportion of our country's coastal areas (inshore waters).

14.6　No subsidies for illegal, unreported and unregulated (IUU) fishing activities will be provided.

14.6.1　Progress by countries in the degree of implementation of international instruments aiming to combat illegal, unreported and unregulated fishing

14.A　Policy guidance and protection of family-oriented small-scale fisheries to sale it's achievement smoothly.

14.A.1　Degree of application of a legal, regulatory policy, institutional framework which recognizes and protects access rights for small-scale fisheries.

14.B　Implement the existing regional and international systems of the United Nations Convention on the Law of the Sea (UNCLOS).

14.B.1　Number of countries making progress in ratifying, accepting and implementing through legal, policy and institutional frameworks, ocean-related instruments that implement international law, as reflected in the United Nations Convention on the Law of the Sea, for the conservation and sustainable use of the oceans and their resources.

| SDGs Goal 15 | Conserve and sustainably use terrestrial ecosystems to ensure the persistence of biodiversity and the prevention of land degradation. |

In 2015, the United Nations launched the Sustainable Development Goals 2030 (Sustainable Development Goals, SDGs) and proposed 17 core goals for global governments and enterprises to jointly move towards sustainable development and address issues such as the gap between rich and poor, climate change and gender parity. -- The third goal is "Conserve and sustainably use terrestrial ecosystems to ensure the persistence of biodiversity and the prevention of land degradation." What are the regulations and core spirit of this goal? Are there examples and reflections inside?

15.1 By 2020, ensure the conservation, restoration and sustainable use of terrestrial and inland freshwater ecosystems and their services, in particular forests, wetlands, mountains and drylands, in line with obligations under international agreements.

    15.1.1 Forest area as a percentage of total land area.

    15.1.2 Proportion of important sites for terrestrial and freshwater biodiversity that are covered by protected areas, by ecosystem type.

15.2 By 2020, promote the implementation of sustainable management of all types of forests, halt deforestation, restore degraded forests and substantially increase afforestation and reforestation globally.

15.2.1　Progress towards sustainable forest management.

15.3　By 2030, combat desertification, restore degraded land and soil, including land affected by desertification, drought and floods, and strive to achieve a land degradation-neutral world.

15.3.1　Proportion of land that is degraded over total land area.

15.4　By 2030, ensure the conservation of mountain ecosystems, including their biodiversity, in order to enhance their capacity to provide benefits that are essential for sustainable development.

15.4.1　Coverage by protected areas of important sites for mountain biodiversity.

15.4.2　Mountain Green Cover Index.

15.5　Take urgent and significant action to reduce the degradation of natural habitats, halt the loss of biodiversity and, by 2020, protect and prevent the extinction of threatened species.

15.5.1　Red List Index.

15.6　Promote fair and equitable sharing of the benefits arising from the utilization of genetic resources and promote appropriate access to such resources, as internationally agreed.

15.6.1　Number of countries that have adopted legislative, administrative and policy frameworks to ensure fair and equitable sharing of benefits.

15.7　Take urgent action to end poaching and trafficking of protected species of flora and fauna and address both demand and supply of illegal wildlife products.

15.7.1　Proportion of traded wildlife that was poached or illicitly trafficked. (Same as 15.C.1)

15.8　By 2020, introduce measures to prevent the introduction and significantly reduce the impact of invasive alien species on land and water ecosystems and control or eradicate the priority species.

15.8.1　Proportion of countries adopting relevant national legislation and adequately resourcing the prevention or control of invasive alien species.

15.9 By 2020, integrate ecosystem and biodiversity values into national and local planning, development processes, poverty reduction strategies and accounts.

    15.9.1 Progress towards national targets established in accordance with Aichi Biodiversity Target 2 of the Strategic Plan for Biodiversity 2011-2020.

15.A Mobilize and significantly increase financial resources from all sources to conserve and sustainably use biodiversity and ecosystems.

    15.A.1 Official development assistance and public expenditure on conservation and sustainable use of biodiversity and ecosystems. (Same as 15.B.1)

15.B Mobilize significant resources from all sources and at all levels to finance sustainable forest management and provide adequate incentives to developing countries to advance such management, including for conservation and reforestation.

    15.B.1 Official development assistance and public expenditure on conservation and sustainable use of biodiversity and ecosystems. (Same as 15.A.1)

15.C Enhance global support for efforts to combat poaching and trafficking of protected species, including by increasing the capacity of local communities to pursue sustainable livelihood opportunities.

    15.C.1 Proportion of traded wildlife that was poached or illicitly trafficked. (Same as 15.7.1)

SDGs Goal 16	Promote a peaceful and diversified society. Ensure judicial equality and build an accountable and inclusive system.

In 2015, the United Nations launched the Sustainable Development Goals 2030 (Sustainable Development Goals, SDGs) and proposed 17 core goals for global governments and enterprises to jointly move towards sustainable development and address issues such as the gap between rich and poor, climate change and gender parity. -- The third goal is "Promote a peaceful and diversified society. Ensure judicial equality and build an accountable and inclusive system." What are the regulations and core spirit of this goal? Are there examples and reflections inside?

16.1 Significantly reduce all forms of violence and related death rates everywhere.

    16.1.1 Number of victims of intentional homicide per 100,000 population. (Disaggregated by sex and age.)

    16.1.2 Conflict-related deaths per 100,000 population. (Disaggregated by sex, age and cause.)

    16.1.3 Proportion of population subjected to physical, psychological or sexual violence in the previous 12 months.

    16.1.4 Proportion of population that feel safe walking alone around the area they live.

16.2   End abuse, exploitation, trafficking and all forms of violence against and torture of children.

     16.2.1   Proportion of children aged 1-17 years who experienced any physical punishment and/or psychological aggression by caregivers in the past month.

     16.2.2   Number of victims of human trafficking per 100,000 population. (Disaggregated by sex, age and form of exploitation.)

     16.2.3   Proportion of young women and men aged 18-29 years who experienced sexual violence by age 18.

16.3   Promote the rule of law at the national and international levels and ensure equal access to justice for all.

     16.3.1   Proportion of victims of violence in the previous 12 months who reported their victimization to competent authorities or other officially recognized conflict resolution mechanisms.

     16.3.2   Unsentenced detainees as a proportion of overall prison population.

16.4   By 2030, significantly reduce illicit financial and arms flows, strengthen the recovery and return of stolen assets and combat all forms of organized crime.

     16.4.1   Total value of inward and outward illicit financial flows. (in current United States dollars)

     16.4.2   Proportion of seized, found or surrendered arms whose illicit origin or context has been traced or established by a competent authority in line with international instruments.

16.5   Substantially reduce corruption and bribery in all their forms.

     16.5.1   Proportion of persons who had at least one contact with a public official and who paid a bribe to a public official, or were asked for a bribe by those public officials, during the previous 12 months.

     16.5.2   Proportion of businesses that had at least one contact with a public official and that paid a bribe to a public official, or were asked for a bribe by those public officials during the previous 12 months.

16.6   Develop effective, accountable and transparent institutions at all levels.

16.6.1 Primary government expenditures as a proportion of original approved budget. (Disaggregated by sector or by budget codes or similar.)

16.6.2 Proportion of the population satisfied with their last experience of public services.

16.7 Ensure responsive, inclusive, participatory and representative decision-making at all levels.

16.7.1 Proportions of positions in public institutions (national and local legislatures, public service, and judiciary) compared to national distributions. (Disaggregated by sex, age, persons with disabilities and population groups.)

16.7.2 Proportion of population who believe decision-making is inclusive and responsive. (Disaggregated by sex, age, disability and population group.)

16.8 Broaden and strengthen the participation of developing countries in the institutions of global governance.

16.8.1 Proportion of members and voting rights of developing countries in international organizations. (Same as 10.6.1)

16.9 By 2030, provide legal identity for all, including birth registration.

16.9.1 Proportion of children under 5 years of age whose births have been registered with a civil authority. (Disaggregated by age.)

16.10 Ensure public access to information and protect fundamental freedoms, in accordance with national legislation and international agreements.

16.10.1 Number of verified cases of killing, kidnapping, enforced disappearance, arbitrary detention and torture of journalists, associated media personnel, trade unionists and human rights advocates in the previous 12 months.

16.10.2 Number of countries that adopt and implement constitutional, statutory and/or policy guarantees for public access to information.

16.A Strengthen relevant national institutions, including through international cooperation, for building capacity at all levels, in particular in developing countries, to prevent violence and combat terrorism and crime.

16.A.1 Existence of independent national human rights institutions in compliance with the Paris Principles.

16.B Promote and enforce non-discriminatory laws and policies for sustainable development.

16.B.1 Proportion of population reporting having personally felt discriminated against or harassed in the previous 12 months on the basis of a ground of discrimination prohibited under international human rights law. (Same as 10.3.1)

| SDGs Goal 17 | Establish diversified partnerships and work together to advance the sustainable vision. |

In 2015, the United Nations launched the Sustainable Development Goals 2030 (Sustainable Development Goals, SDGs) and proposed 17 core goals for global governments and enterprises to jointly move towards sustainable development and address issues such as the gap between rich and poor, climate change and gender parity. -- The third goal is "Establish diversified partnerships and work together to advance the sustainable vision." What are the regulations and core spirit of this goal? Are there examples and reflections inside?

17.1 Handle the transfer, popularization and dissemination of environmentally friendly technology to improve energy efficiency, reduce pollution, and promote waste recycling.

    17.1.1 Handle the total amount of transfer, popularization and dissemination of friendly environmental technology.

17.2 Promote medical cooperation plans, assist special types of countries (low-developed countries, small island countries, and African countries) to train medical personnel in Taiwan and provide scholarships for scholarships to accept public health medical disciplines (medicine, nursing, pharmacy, etc.) related majors in Taiwan training.

    17.2.1 The number of assisting in training the number of foreign medical personnel.

    17.2.2 The results of the implementation of various medical cooperation plans with diplomatic countries.

17.3 Continue to provide Taiwan scholarships to outstanding students from countries with diplomatic relations (and some developing countries) to study in Taiwan.

    17.3.1 The number of approved for Taiwan scholarships.

17.4 Continue to assist in the promotion of local water and sanitation related projects in developing countries.

    17.4.1 Assist in promoting water and sanitation related projects in developing countries.

17.5 Handle various technical assistance plans for trade aid types.

    17.5.1 The total amount of technical assistance plans for trade assistance type. (NTD)

17.6 For developing countries, continue to use our country's advantages to assist their development. In accordance with the relevant agreements of the World Trade Organization (WTO), special and differential treatment will be granted to such countries, and it will be discussed to increase my country's preferential treatment of "duty-free and quota-free" for low-developed countries (LDCs).

    17.6.1 Our country provides low-developed countries with tariff-free preferential treatment tax as a percentage of my country's customs import tariffs.

17.6.2 The total number of foreign technical cooperation projects.

17.7 Continue to implement extraordinary poverty eradication programs in accordance with the needs of the international community.

17.7.1 The ratio of government resources directly invested in poverty eradication programs.

17.8 Actively participate in WTO trade and environment discussions and negotiations, strengthen mutual support between trade and environment, and promote a universal, standardized, open, non-discriminatory and fair multilateral trading system.

17.8.1 The tax rate of the list of goods attached to the environmental commodity agreement.

17.9 Utilize bilateral and multilateral environmental protection cooperation projects, and use technology to assist energy construction or public and private sectors and non-governmental organizations to work together to improve environmental management and pollution prevention in developing countries.

17.9.1 The number of planned activities performed.

17.9.2 The total number of countries participating in the project.

17.10 Continue to cooperate with Indonesia, Vietnam, Thailand, and develop cooperation with India and other countries to select elites to study in Taiwan to promote international teacher training cooperation.

17.10.1 Number of elite trainings in cooperation with developing countries.

Missing Corner of SDGs   Digital technology serves mankind and the planet.

2030 is the acceptance schedule of the United Nations Sustainable Development Goals (17 SDGs). As time approaches, countries around the world are preparing for mid-term assessments. At the same time, the public is also very curious about what the world will become after the COVID-19 epidemic. Although there are many uncertain factors, one thing is certain that our lifestyle and more and more economic transactions and government affairs will be transferred to the Internet. In 2015, 193 countries around the world committed themselves to achieving 17 SDGs by formulating a transformative sustainable development program that connects human health, prosperity, the environment, and equality. This comprehensive and sustainable program is of vital importance to people around the world.

However, the coronavirus crisis in early 2020 has greatly highlighted the incompleteness of the current 17 SDGs. Today's existing 17 SDGs have not yet made appropriate responses and considerations to the powerful force that gives humanity a new definition of the future-the "Digital Age".

Current status: Digital gap accelerates social risks and uncertainties

The digital generation is rebuilding the old social structure, and is driving changes overwhelmingly at an unprecedented scale and speed. Decades ago, the philosopher and journalist Marshall McLuhan put forward the motto: "Humans create tools, and in the end tools also create us." Indeed, today's digital technology is redefining reading, consumption, voting, and the way people interact with each other. Many risks and uncertainties have also emerged, including threats to individual rights, social equality and democracy, and these risks and uncertainties have been amplified by the "digital divide". The digital gap refers to the degree of internet penetration, and the difference in opportunities and abilities to use digital information or products, which leads to a greater gap between the rich and the poor.

Our initiative: SDG Goal 18 - Digital technology for the well-being of mankind and the planet.

Along with huge risks of harm, comes huge opportunities. We believe that we can make good use of the wider influence of the digital generation to guide society toward a common SDGs, including achieving zero carbon emissions and a more equal society. In the digital generation, through technology, we can decentralize power and transfer it to other stakeholders who need it. At the same time, we can shift social consumption habits to low-carbon footprint products. We can also shift social mindset from using fossil fuels to using renewable energy. Therefore, we need a new sustainable development goal for the digital generation -- SDG18, to make good use of this powerful transformative force to benefit mankind and the planet.

Internationalization has always been a common standard and common language necessary for conversation and run business between countries. The country's economic lifeline is endless. Taiwan's corporate CSR performance has reached world-class standards, but it is showing an M-shaped development. more than 80% of SMEs have yet to CSR concept and not carried out or not implemented CSR issues, TransAct understand that Taiwan's traditional concepts are not good at expressing and do story marketing, but actually each company in its original behavior and actions are already performing CSR. Now we are starting to assist more SMEs through the TransAct's SESP platform, using

digital technology to let SMEs maximize CSR by using a small amount of resources, and this is also how TransAct implement CSR's 18th target: the correct use of digital technology to promote the core value of human peace and well-being, through the accumulation of digital knowledge and technological wisdom, is an important common asset of mankind, and it can indeed be continuously updated. At the same time, combining artificial intelligence, empathetic and intentional human wisdom and social wisdom, using the potential of digitization and related new knowledge wealth to deal with 21st century's major mankind challenges, so the education system and global knowledge transfer can be fundamentally transformed. Three well-known research institutions in the field of sustainable development, including the Austrian Institute for Applied Systems Analysis, the Stockholm Resilience Research Center, and the United Nations Sustainable Development Research Network, call for the world to use the United Nations "2030 Agenda for Sustainable Development" as the basis for immediate key transformation actions. And in 2019, they discuss about how to accelerate the implementation of key transformation aspects through digital tools with the topic of "The Digital Revolution and

Sustainable Development: Opportunities and Challenges." According to the Secretary-General of the United Nations through the further development of a series of round-table discussions with governments, the private sector, civil society, international organizations, academic institutions, technical communities and other key stakeholders, the United Nations envisage to take the following eight big action:

1. Achieve global connectivity in 2030.
2. Promote digital public goods to create a fairer world.
3. Ensure that all people, including the most vulnerable can enjoy digital inclusion.
4. Strengthen digital capacity building.
5. Ensuring the protection of human rights in the digital age.
6. Support global artificial intelligence cooperation.
7. Promote trust and safety in the digital environment.
8. Build a more efficient digital cooperation structure.

The international trend makes TransAct expect that through the technology system, power can be collectively decentralized and transferred to other necessary stakeholders; this digital revolution: including virtual reality and augmented reality, 3D printing (Additive manufacturing), general artificial intelligence, deep learning, robotics, big data, IoT and automatic decision-making systems that have entered the stage of public debate in many countries. Looking back on the past, it is hard to believe that the "2030 Agenda for Sustainable Development" or the "Paris Agreement" rarely mentions digitalization. It is obvious that digital changes are becoming a key driver of social transformation. Therefore, many enterprise, governments, civic groups, and academia jointly issued a statement in June 2020, "The Montreal Statement on Sustainability in the Digital Age", laid the foundation for the initiative of SDG 18. The declaration hopes the society will understand that tackling the climate crisis, building sustainable global development, and realizing an equal digital future are the issues that should be endeavored and are interrelated. Towards sustainable transformation must reconcile their threats, opportunities, and the dynamics of the digital revolution while the digital transition completely change all the aspects of global society and economy, thus changing the interpretation of a model of sustainability itself.

Digitization is a cross-scale economic, social and cultural connection that combines a powerful multiplier of real and virtual situations. More importantly, it creates characteristics of technological systems (such as artificial intelligence and deep learning) that strengthen human perception and have cognitive capabilities. At least in some functional areas, it will eventually complement, perhaps even replace, or ultimately far exceed human cognitive abilities. Digitization is no longer just a "tool" to solve sustainability challenges. It is also an important driving force for destructive and multi-scale changes. In order to sustain important issues in the future, digitization also needs to respond to the greatest impact. With the smallest opportunities and minimal risks, six important mechanisms have been proposed internationally to link digital power and sustainable strategies:

1. Create a sustainable digital perspective in the science, research, and R&D communities, and change the vision and mode of innovation.

2. Set appropriate prices to drive market forces, such as carbon pricing and ecological tax reforms, to stimulate digital innovation actions to support sustainable solutions.

3. Use digitization to concretize and formulate a transformation roadmap, including clear definition of goals and milestones for energy, transportation, land use systems, urban and industrial sectors, and help the market and planning process move to a more sustainable direction.

4. Invest in digital modernization projects at the national level to increase the digital knowledge in public institutions on a large scale, so as to build the governance capabilities of the digital anthropology.

5. By supporting and expanding a strong network with the digital research community, establish transformative and sustainable research.

6. Establish a dialogue mechanism with the private sector, civil society, science, and the state to jointly discuss the development of the digital anthropological system, society, and normative guarantees.

Digitization will be the reality and challenge of the new economy, society and culture. Virtual reality, artificial intelligence, deep learning, big data, and more and more activities used in planning and creating situational processes will improve the cognitive ability on understand the meaning of decision-making, knowledge breakthroughs and explosions will bring unprecedented new potential to mankind to our complex social ecosystem. Decision makers, researchers, enterprises and civil society actors must strengthen their efforts to understand and explain the multiple impacts and far-reaching effects of the structural changes in the industry because of the digital changes. The use of science and technology as a kind of fire that triggers investment in various fields can expand the spillover effect of science and technology policies, so that it can create a foundation for shaping the digitalization process, appropriate digitalization processes and related technologies (such as new composite materials, Nanotechnology, nanobiotechnology, genetic engineering, synthetic biology, bionics, quantum computing, 3D printing, and human-enhancement, technology augmentation) will transform Homo sapiens into Homo digitalis. Artificial intelligence, deep learning and big data will change science and open up new doors for the next stage of human civilization. Virtual access to the most advanced global knowledge about humans and the earth can achieve fair, justice and safe in future. Digital power can promote cultural, institutional, and behavioral innovation. Transnational communication networks can help establish a global networked society, transnational governance mechanisms, global common wealth perspectives, a culture of global cooperation, and transnational identity, and may also create new (secondary) Culture, people from all over the world in the virtual network, may improve our understanding of cultural diversity. Virtual reality allows humans to visit, understand, enjoy and experience the global ecosystem without having to travel a long distance. At the same time, through the digital voting process, new options for promoting democratic development mechanisms are rapidly expanding, this includes online checking whether local decision-making involves practical but important governance issues related to transformation and reform preferences. In summary, the convergence of digital technologies will definitely enhance human functions and cognitive abilities, confirming that human functions have been greatly improved in the last century. Unprecedented achievements have been made in medical and health, competitive sports, and knowledge. Over the centuries, human life spans have multiplied. With the increase of digital, the use of artificial organs and prosthetic limbs will also begin to experience major breakthroughs, such as the new improvement of exoskeleton reinforcement and human function enhancement, and the ancient science fiction of machine-made machines. The vision has come true, freeing people from physical labor, enhancing and enhancing cognition and physical fitness. These emerging innovations show the potential positiveness of the new era of mankind. Digitization provides excellent possibilities and makes it a sustainable transformation and development.

c. 2021 is the first year of ESG development, it brings opportunities and hope

2021 is the first year of ESG development. The country and industry face up to its importance and bring about opportunities and hopes. Initiating changes and new opportunities has become the top priority. The "2021 Global Views Monthly Magazine ESG Sustainable Finance Forum" had an in-depth discussion of the sustainability issues pursued in 2021, and the key to the future of financial development. Yang, Ma-Li, the president of *Global Views Monthly* Magazine, pointed out that under the promotion of the Financial Examination Bureau, (FSC)'s policies, ESG issues will be regarded as a prominent study in 2020, and 2021 will be the first year of ESG development. All industries are actively engaged in this issue and are developing vigorously. The current digital challenge is to solve the arduous sustainability issues of the digital generation and reverse it. Reversing the situation has become a pressing matter of the moment. Urgent needs in the perspective of sustainable development agenda of the 2030, 2030 remaining less than 10 years away, society and its government is in the most important critical period, we understand and shape the governance towards sustainable digital revolution is not cure-all, because the future is inherently uncertain, time is a very precious and scarce resource and we must use it wisely. We can fully mobilize in less than 10 years, and by using digital opportunities to create sustainable of society, and to learn how to manage the number of bits of green economy and a stable, fair and open digital society, positive use of social impact of digital technology and artificial intelligence, the integration of virtual and real space and real circumstance, avoid further weakening social cohesion. The challenge in the future will undoubtedly be cognitive enhancement. This challenge lies in the establishment of a resilient, adaptable, knowledgeable and inclusive "responsible society".

The potential of digitalization to make significant progress in education, health, equity, and prosperity is undeniable. The way and place of life and work, how to use the leisure time that increased, and how to interact with the immediate, local, and broader community's members, also understand emerging challenges such as artificial intelligence, automatic decision-making systems, and virtual spaces. Significant changes in these methods will bring social impact. TransAct's platform hopes to bring togetherness of the views of the digital and sustainable research community, and work with international partners to take advantage of the opportunities of digitization, virtual reality, and artificial intelligence to curb potential risks and link digital and sustainable transformation. Following the basis, an interdependent system architecture has been created to help the reconciliation of digital management and sustainable transformation:

1. Education: People need to be able to understand and shape the emerging digital shifts.

2. Science: The new knowledge network must create transformative knowledge to integrate digital and sustainable-oriented transformations, avoid approaching digital critical points, and establish a normative framework that integrates mankind and smart machines era.

3. Promote the modernization of the country: Public institutions are not yet ready to understand and govern digital dynamics. But large-scale modernization and education programs are necessary.

4. Experimental area: Learning and applying by doing, especially in the early stage of innovation, are the main principles of technology and system communication. Some creative area should be built to foster rapid learning, and include the possibility of outputting "crazy ideas and startups".

5. Global governance: The digital revolution has global influence in building alliances. For example, the modernization of the United Nations will be shaped by the digital age.

6. "New Humanism" (WBGU, 2019): The "2030 Agenda for Sustainable Development" can be regarded as a new "social contract" to the world. It changes our values and visions for the future beyond 2030, and moves towards the direction of overall sustainability. This also means that mankind and the earth have new normative goals for the future, a

new development model that decouples materialism, the environment and the negative externalities of the earth system, and a new normative guarantee for everyone.

COVID-19 epidemic has put the world at a crossroads. SDGs should be regarded as the key to achieve sustainable development in 2050 and beyond. SMEs are the mainly enterprises in Taiwan, while facing transformation upgrading, lack of digital talents and software resources, upstream and downstream information, international market development issues, big data, AI, 5G, IOT and other new technology that have not been integrated; make the transition towards sustainable development requires a long time to finish. It is urgent to have a corresponding management policies, reward programs, and value changes, but the facts show that the world only exists in a few countries. In order to achieve these six key changes, there are six important mechanisms that are promoting to link digital power and sustainable strategies.

This global epidemic has exposed many problems arising from the world's high dependence on the economy. For companies, this epidemic is not only a crisis but also a turning point, and it has brought normal changes to sustainable development. Sustainable investment has become an irreversible trend, CSR methods have changed from temporary, incremental, and transactional methods to strategic, social purpose-driven and transformational models. CSR is shifting from the driving force of "frugal" to the motivation "indispensable for business success". The digital technology platform of TransAct makes each specific partnership unique and requires partners to promote innovation and success. They have the same thing: CSR commitment is in line with the company's basic values and strategic direction, so it is trustworthy and real.

According to McKinsey research, the digital economy will contribute approximately 49% of global GDP by 2025. Taiwan's digital economy's GDP ratio in 2025 is targeted at 29.9%. With reference to the digital economic development speed of major economies, it is estimated that Taiwan's digital economy's GDP ratio in 2030 will be close to 40%. Although sustainable investment has been talked for more than ten years, but the real prosperity comes recently. In the entire investment industry value chain, stakeholders include: customers, sales, partners, employees, shareholders, invested companies, suppliers, regulatory agencies, rating agencies, media, local communities, non-governmental organizations and other civil organizations. The corporate value driven by this trend is invisibly being reshaped, and sustainable investment has become a key catalyst to help enterprise enhance their value. A knockout competition for companies that expands globally and influences future competitiveness has quietly debuted. The ticket is ESG. If companies do not have ESG, they may not even have the qualifications to enter the competition and compete with the same industry. Studies have pointed out that companies not only have a lot of discussion about the epidemic in the financial report meeting, but also have a significant increase in non-financial reports. ESG is as important as EPS. It must be implemented from senior executive to entry level employee, from internal to external, and the company is also repositioning their own business models, review and control risk factors; from the earth to this land, and even for the diversity and tolerance, health and safety, compensation and benefits, labor relations and equal rights with employees and so on, therefore; TransAct's digital technology platform spreads and shares the purpose of the government, and assists in the promotion of employment-related subsidies from the Ministry of Labor (youth, workplace readjustment, employment incentives, middle-aged and senior citizens, secure employment, job redesign, industry-university cooperation). Revitalize and increase the employment population, corporate human resource improvement plan subsidies to promote employee function development and work efficiency, work-life balance subsidies, improve the working environment and welfare of employees, greatly reduce employee turnover rate and increase work efficiency. In addition to promote the integrity and loyalty between company and its employees, it also imprints its corporate image and unites its employees' centripetal force. As well as assisting companies in constructing a talent training system like TTQS Enterprise Edition and Institutional Edition, obtaining national licenses, and even planning iCAP function-oriented courses based on the characteristics and levels of corporate talents, they are all transformed into CSR based on the core development of the enterprise, motivating enterprises to respond to market challenges It will also be more obvious in the long run. Under the influence of the COVID-19 epidemic, mankind has a deeper understanding that ESG is an important key to a sustainable future. From individuals to enterprises and countries should all stand up and take actions to reverse the crisis and to make the world better.

Taking the estimated data of the G20 summit as an example, if the United Nations permanent development goal is to be completed by 2030, an additional US$ 2.9 trillion must be invested on the basis of sustainable energy; looking at the entire capital market, ESG has been included in the basic investment projects, based on the forecast of investment banking Bank of America (BofA), ESG fund assets under management will reach more than $ 20 trillion in the next 20 years, Union Bank of Switzerland even estimated that ESG fund assets under management will reach $ 1.5 trillion in 2036. Focusing on investment performance in ESG: BNP Paribas Funds Energy Transition increased by 174% within six months, and O-Bank launched Taiwan's first "influence of deposit" project, financial management can accumulate benefits by CSR. In addition, the ESG disclosure in December 2020: Taiwan's ESG ranks the highest in Asia, and Taiwan government takes the lead in the officially calculating of the "green procurement" score in March 2021. Therefore, if a company wants to be favored by capital market finance on ESG and gain more cooperation and procurement opportunities, it is bound to internalize CSR/ESGs as a company's innovation and growth strategy. However, CSR actions with different industries and cultures make SMEs confused about the content and projects they implement. This once again proves that our society needs corporate solutions, TransAct's digital platform will drive and induce companies to play a greater energy, provide customized enterprise innovative business model to find a new opportunity by using practical experience. Implement business policies, plan specific CSR strategies, and have an in-depth understanding of the facts from the process of understanding to implementing CSR for all stakeholders. Through in-depth communication with the enterprise and fully implementing the CSR plan. Obtain subsidies from various government ministries, science and technology project plans, and rescue guidance, allowing companies to tide over difficulties through government policies, attach importance to CSR, advocate national policies with action execution, and assist companies with epidemic relief subsidies and relief loans, and youth Entrepreneurship and micro-Phoenix and national investment "Taiwan three major investment programs" investment in Taiwan office and other assistance. Raise the threshold of competition within the same industry; let a fulcrum drive the international economic benefits additional, and actively create an innovative value chain industrial environment. As Apple Inc. announced to reach 100% of carbon neutrality in 2030, many Taiwan manufacturers are inside Apple's supply chain, facing the explosive growth of the green supply chain in the next 10 years, the materials used in daily necessities, from clothes, glasses to shoes, will undergo tremendous changes, and the entire industry supply chain will be transformed and upgraded to a circular and sustainable system. Taiwan has great advantages and business opportunities in circular economy and sustainable innovation. TransAct hopes that Taiwan's international competitiveness and Taiwan's foreign relations can move towards a new generation of new national strength.

Reference

- CommonWealth Magazine (Editorial Department of Future City). SDGs 1~17. https://topic.cw.com.tw/event/2020sdgs/

- Editorial Department of Future City. (2021, January 27). About SDGs: What is The United Nations Sustainable Development Goals? Take a Look at the 17 Goals. https://futurecity.cw.com.tw/article/1867

- Zhu-Yuan Zhu. (2019, May 16). [Column] Zhu-Yuan Zhu: Sustainability Should Not Have a Threshold. https://csr.cw.com.tw/article/40963

- Xin-Ning Li (translator). (2020, October 14). The Montreal Digital GenerationSustainability Protocol. https://rspre.ntu.edu.tw/zh-tw/m01-3/understand-risksociety/295-sustainable-tra/1488-1014-systemdynamic.html

- Qian-Ci Lin (translator). Ying-Ting Guo (proofreader). (2019, November 13). *The TWI2050 Report (2019) & Highlights from The Digital Revolution and Sustainable Development: Opportunities and Challenges.* https://rspre.ntu.edu.tw/zh-tw/m01-3/tech-pros/1319-1081113-key-factor.html

- The American Institute in Taiwan. Descriptions on the SDGs. https://www.ait.org.tw/wp-content/uploads/sites/269/un-sdg.pdf

- The Storm Media. (2020, August 27). Why Does the Global Capital Market Value ESG Increasingly? Investment Experts: Numbers Don't Lie. https://www.storm.mg/article/2980627

- Qing-Xiang Chen. (2014, July). How to Practice CSR Comprehensively. https://www2.deloitte.com/tw/tc/pages/risk/articles/newsletter-07-6.html

- Xiao-Wen Huang. (2021, March). The New Look ofSustainable Management from the Perspective of Finance, Shareholders, and the Capital Market. https://www.accounting.org.tw/flptopic.aspx?f=306

- Yan-Jing Gu. (2021, January 29). Investing in Sustainable Development Together is the Consensus to Prevent Potential Risks. https://www.gvm.com.tw/article/77561

- Amy Luers (author). Yu-Hong Zheng from CSRone (translator). (2020, November 22). The Missing SDG: Ensure the Digital Age Supports People, Planet, Prosperity & Peace. https://csr.cw.com.tw/article/41743

- Apple. (2020, July 21). Apple Commits to be 100 Percent Carbon Neutral for its Supply Chain and Products by 2030. https://www.apple.com/tw/newsroom/2020/07/apple-commits-to-be-100-percent-carbon-neutral-for-its-supply-chain-and-products-by-2030/

- NTU SDGs. (2020, January 2). The World in 2050: Transformations to Achieve the Sustainable Development Goals. https://oir.ntu.edu.tw/ntusdg/outcome/%E6%83%B3%E5%83%8F2050-%E8%90%BD%E5%AF%A6%E6%B0%B8%E7%BA%8C%E7%99%9D%BC%E5%B1%95%E7%9B%AE%E6%A8%99%E7%9A%84%E9%97%9C%E9%8D%B5%E8%B-D%89%E5%9E%8B%E8%A1%8C%E5%8B%95/

- The United Nations Statistics Division. SDG Indicators. https://unstats.un.org/sdgs/metadata/

D. CSR practical action

a. University Social Responsibility (USR)

The importance of university social responsibility for sustainable development of Taiwan industry

Chairman of National Kaohsiung University of Hospitality and Tourism

Professor Rong-Da Liang

While CSR is being put into practice from companies to corners of the society, universities have also started to promote USR, hoping to bring change to the society through its specialty and advantages with young students. In recent years, the Ministry of Education has been actively promoting the practice of USR, hoping to lead universities out of the traditional frame of teaching and learning, increase the interactions between universities and local communities, and cultivate talents that meet the needs of the development of society. These could help develop the features of each university, understand social responsibility, and meanwhile solve problems in each corner of the society. The roots of these problems may be difficulties in marketing, a lack of professional competence, or a lack of characteristics, and these are abilities that universities have. If professional academic teachers and motivated, creative students could do their best to combine resources of CSR, have the concept of CSR and USR, and start from the two aspects of corporation and education, the results and influence would be impressive.

Many companies consider CSR simply a way to give away money like a donation campaign, yet that is not true. The aspects of CSR are widely diverse; for example: giving employees a decent working environment, or assisting neighbors with their regional features. More importantly, a company could use its specialties to build something unique and durable, as in combining its core capability. The structure of CSR can be divided into three stages: first, the commencing stage, which is the stage to drive CSR from the inside and outside. The motivation to start it comes from corporate social norms, government policies, leadership value, and organizational culture. Second, the company starts implementing CSR, using: (1) economic strategies like charity and sustainable tourism products, (2) environmental strategies like energy saving, environmental protection, and sustainable resources management, (3) social strategies like customer education and industry-academia collaboration, (4) cultural strategies like sustainable cultural heritage and assistance in cultural and creative industries. Third, the positive feedback stage, in which the company may see some results like positive feedbacks, a rise in corporate image, and inner or outer customers' satisfaction or pursue of sustainability. Among the social strategies, industry-academia collaboration is the main subject of this article. I will share my experience as the host and co-host of university social responsibility in the MOE, along with my results in promoting the practice of USR in the National Kaohsiung University of Hospitality and Tourism. I will show how to actively help community associations build community features, aid local communities and companies to co-develop from the academic view, create a win-win situation, and achieve the goal of sustainability together.

The NKUHT has always been known for cultivating talents in hospitality and tourism. It is also the key partner of hospitality and tourism industries. The university has been strongly promoting students' and teachers' implementation, creative businesses, local creation, and internationalizing industries. With the collaboration of Chen Jen Kuen, the district executive of the Qiaotou District of Kaohsiung, we have connected with the National Kaohsiung Normal University, the National Kaohsiung University of Science and Technology, and the Chaoyang University of Technology, and received the MOE's USR implementation subsidy in the 106th year of the Republic Era. The core values of the subsidy program are to apply the creative specialty of the NKUHT, to implement the economic, marketing, and agricultural specialty of each university, combining the power of local communities and industries, and promoting the local dairy farming, honey, and fruit industries' characteristics. Members of the program think that building sceneries and ingredients with local features by combining the power of each field, and helping locals carry out unique agricultural specialty and local travel itinerary would solve problems regarding the regional economic development of the Qiaotou District. Firstly, they would like to develop agricultural specialties, assist organic development, and help young farmers with marketing. Secondly, they would increase media exposure of Qiaotou.

The long-term goal is to build unique style and features for the Qiaotou community, including transforming local ingredients to local souvenirs, transforming local agriculture to agricultural recreation, developing regional transformation and features, and since making Qiaotou "the sweet garden" of Kaohsiung. The local residents had missed out on opportunities to increase added value because their management mode was mostly production oriented. Therefore, the team has planned special souvenirs and community travel itinerary, hoping to solve the core problem. The team has combined milk, honey, and dragon fruit, and developed a series of dragon fruit jam, marshmallow biscuits, sparkling juice, and ice cream as the special foods of the ShueiLiou Farm of Qiaotou. After the implementation of the program, firstly, with the efforts of the members and students, they have helped reinforce economic and local features development to solve the problem of lack of local resources. Secondly, after multiple interactions and visitations, the community has formed understanding with each other. The program has offered practical creative thinking and assisted in development of community activities and environmental protection, which have solved the problem of lack of manpower and resources. Thirdly, for local industries, the program has utilized the special foods developed from local ingredients to solve the problem of excessive ingredients, meanwhile building the base for regional features. A USR program like this is not only universities using its specialties to help the community, but also a way for the community to showcase its practice for the education in universities, and to solve the problem of lack of motivation in students.

Similar to the idea of the government urging universities to implement social responsibilities, I believe we can reinforce the collaboration between universities and local community development, and let universities out of their academic ivory tower. In the process of regional creative development, we can put in more academic energy in developmental issues like local industries, community cultures, urban and rural, etc., making universities active contributors to local sustainability, and help the Taiwanese society to grow together. Since the promotion of USR by the government, it has developed into two types of issues: "local creation" and "international connection". In order to connect with the world and steadily participate in the globe, the goals are to stimulate and recreate industries and attract population. It is also due to this generation of infinite possibilities, it can drive USR to bring motivation for the society to continually change for sustainability. During this moment that the influence of CSR has been expanded in the public eye, the rise of USR is the fuel on the way of pursuing sustainability. My experience, practices, and results sharing can be the best teaching material for more improvements of the society and students!

b. Global Action

Climate Change and Environmental Sustainability of "Climate Action"

CEO of Weplus High-Tech Corporation

Xue-Hui Wang

Air is an essential factory in a human being's survival, yet according to statistics of the WHO, about seven million people in the world have died because of air pollution indoors and outdoors. Air pollution has become the top of environmental risks in both urban and rural areas. Although the public has become substantially more aware of public health crisis and has been trying to implement it, we must understand more about the deep influence air pollution has on the earth and human beings. According to research, the harm air pollution brings has already exceeded personal health, and has extended to issues of climate change, clean water sources, renewable energy, and agricultural development.

"Sunlight, air, water" are the three factors for human basic survival. However, the amount of polluted air now can easily damage every organ and cell of a human from head to toe. Take an adult human for example, they inhale and exhale seven to eight litters of air every minute in average, which equals to breathing 11,000 litters of air every day, no matter if it is clean or not. We all know that air pollution damages the faculty of the lungs and the windpipe, and accelerate the speed of death. Nonetheless, according to what researchers of Forum of International Respiratory Societies indicated in the magazine *Chest,* after intensely polluted air is breathed in the

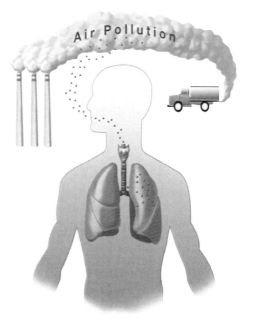

body, the areas it affects are much wider. Aside from cardiopulmonary vascular disease, it could also result to diseases like diabetes, dementia, bladder cancer, osteoporosis, reduced fertility, etc. From the fact that lung cancer has continually been the first or second place in the Taiwanese cancer ranking for 40 years, we could see that air pollution has already been harming human health without us noticing. Yet, in the pandemic started in 2020, the CDC stated in the COVID-19 situation that the virus could float in the air for hours to spread, admitting that potential problems proposed by public health experts of the virus being spread through air indeed exist.

During this time of pandemic, with the effort of Professor Wong from the Resources and Environment Information Development Center of Weplus High-Tech Corporation, "Air Exchange Natural Transform System" has played an important part. It is the first in the world to create air-segregating energy transform technology, adopting rust-proofing antiseptic division plate to completely divide the pathway of intake and exhaust. While it is transforming energy, it performs sterilization, thoroughly resolves the problem of cross-contamination, and reduces the chances of the pandemic spreading wider. It is applicable to the medical field, and you can set it to "negative pressure environment" or "positive pressure environment" depends on your needs. It is in use in Show Chwan Memorial Hospital and Min-Sheng General Hospital as "the best tool for epidemic prevention". Furthermore, this system transforms energy through the pressure difference between the water vapor molecule inside and outside of the building, which on one hand effectively reduces the concentration of carbon dioxide, and on the other hand reduces the speed of global warming. Besides preventing cross-contamination in the air to prevent the spread of the epidemic, during the transformation of energy, it does not emit hot air, and it reduces the concentration of carbon dioxide, providing the building the lowest energy consumption. Additionally, it creates a nature-like indoors environment to slow down the speed of global warming. In 18 CSR indicators, the technology has made Taiwan known by the world, and benefitted lives and environment. The enterprise tries hard to implement the climate act of the 13th indicator in the SDGs, and put effort into the goal in 2050 with TransAct.

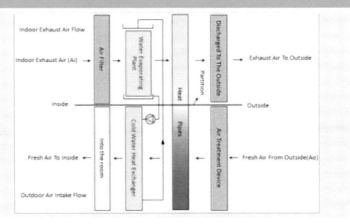

Air segregation material: rust-proofing antiseptic division plate

It uses a rust-proofing antiseptic division plate to completely divide the pathway of intake and exhaust, solving the problem of cross-contamination in energy transformation.

c. Global Ocean Treaty

Greenpeace Organization - Global Ocean Treaty

Fishery

Chairman Tian-Yin Lin

The ocean is the mother of lives, the cradle of beings. No matter in the Evolutionary theory, or in the Age of discovery, the important role the ocean had played in the history of human development is beyond words. To mankind, the ocean is valuable not only because it can adjustment global climates, which creates the environment human live by, but also because of its rich living resources which makes it an important food source of human beings. The ocean is where lives are born and nurtured, it covers 71% of the earth surface area, 99% of the species habitats, providing and cycling on this earth. It is not only the reason the circle of live continuously, but also a significant role in the evolution of human culture.

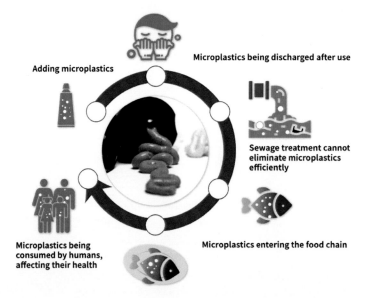

Adding microplastics

Microplastics being discharged after use

Sewage treatment cannot eliminate microplastics efficiently

Microplastics being consumed by humans, affecting their health

Microplastics entering the food chain

Taiwan considers itself a maritime power, but the government pays few efforts for the sea. The two mainly aquatic ecosystems are marine ecosystem and freshwater ecosystem, while only 41% of world-famous fish types habitat in freshwater ecosystem. Taiwanese people know too little about the sea, to most people, the ocean is a total stranger. The very first cross season marine investigation in history, an exploratory research through four seasons, had shown its result: the ocean is crying for help. The result tells the tragical

Reference: The Ocean Cleanup.

truth of how the ocean is occupied by plastic wastes. According to American Oceanographer Jenna Jambeck's calculation, around 8 million tons of plastic wastes are discharged into the sea every year, bringing existential crisis to almost 700 species such as sea turtles, whales or birds. Plastic particles degraded from plastic products adsorb heavy metal or Dioxin easily, and as the marine pollutants increase, the particles are more likely to be miseaten by the marine animals and thus become one part of our food chain. Taiwan Ocean Cleanup Alliance (TOCA) once cooperated with international scholars, analyzing the results of 541 beach cleanup activities hold by Taiwanese people for the past 12 years. The result shows that 90.8% of the garbage collected from beaches of Taiwan are totally or partial made of plastic or plastic. The top 5 common seen trashes are: 1. Plastic bags, 2. Plastic bottle caps, 3. Disposable tableware, 4. Fishing equipment and 5. Straws. The Kuroshio Ocean Education Foundation collects plastic waste using American NGO 5 Gyres' method, and further explains the collected plastic waste by dividing it into the following five categories: 1. Rigid plastic, 2. Soft plastic, 3. Foam plastic, 4. Plastic fiber, and 5. Round plastic particle.

According to British research, there is 12 million 700 thousand tons of garbage discharged into the ocean, 94% of it sinks into the sea bottom while the rest 6% goes to the beaches. The World Economic Forum has warned, if we don't change the limitation of plastic waste, the amount of plastic waste in the ocean will surpass the amount of the fish by 2050. Environment consultancies and domestic marine scholars together announced their research on undersea garbage in Taiwan, revealing that the density of undersea garbage in the 8 main offshore of the west coast is 1.5 times higher than global average, and more than 90% of the garbage are fiber cloth and film plastic. Undersea garbage is seldom concerned in the past, most people take more care about the visible garbage on the ocean surface or beaches. Scholars have warned that undersea garbage not only pollutes the ecosystem, but may be eaten by us human through the food chain once it gets into the bodies of fishes or planktons.

11 waste plastic bottles can be made into a new pair of sneakers

7-8 waste plastic bottles can be made into a new T-shirt

1-1.5 waste plastic bottles can be made into a new plastic bottle

Picking up garbage and doing charity simultaneously to help more people, is really an act of killing two birds with one stone. The amount of plastic waste in the ocean is now over 1.5 billion. If we don't take actions now, the amount will be twice as large as it is now by 2030, 10 years from now. By 2025, every 3 tons of fish will contain 1 ton of plastic, and in 2050, there will be more plastic in the ocean than fish (by weight). The whole world pays high attention to ocean pollution. The coastline of Taoyuan is 46 kilograms long, by the end of September, 2000, Taoyuan City Government had held 71 beach cleanup events along its coastline, removing 218 tons of garbage in total. Taoyuan City Government also work together with local producers, making ecofriendly shoes and sport shirts with recycled plastic bottles. 2020, they even discuss about purchasing discarded nets from the fishermen, and launched the driftwood reuse demonstration program. They collect driftwoods, making them into recycled furniture or plank road surfaces, hoping to reuse the marine debris so as to reduce the waste. The Hotai TOYOTA Corporate held a beach cleanup activity on the morning of April 18, 2020, gathering more than 10 thousand people to go pick up garbage and trashed bottles at Guanyin Coastal Recreation Area, Guanyin district, Taoyuan City. Taoyuan City mayor Cheng Wen-tsan said, Hotai Motor once again holds beach cleanup activities at 17 different places all around Taiwan, and will donate 10 NT dollars for each discarded bottle picked up to Taiwan Coral Island Association as an environmental education for elementary students across Taiwan, implementing their corporate social responsibility, and setting an example of green enterprise. According to "The Time Has Come: The KPMG Survey of Sustainability Reporting 2020," over 60% of enterprise CEOs in both global (63%) and Asia Pacific (66%) state that under the changing environment circumstances, they will value corporate social responsibility and the ESG issue more, and more than half (52%) CEOs in Taiwan as well think that the pandemic emphasis the importance of the ESG issue. How do the enterprises get out from the dilemma and break new ground in the coming new year, pursuing corporate sustainability will be a major issue of many companies.

To do so, I want to represent the fishery, wishing us can make good changes for the marine ecosystem of Taiwan, and hoping to implement corporate social responsibility starting from southern Taiwan. Me who do business in Japan wants to share some practice of Shodoshima, a small island of Japan with everyone. This uncelebrated Shodoshima, Japan is enriched with the grace of the sea. It was a vital sea transportation route of Japan nearly a century ago. Facing the impact of industrialization and globalization brought in modern days, the islands in Setouchi region suffered from rural flight and population ageing, Shodoshima thus became an island with no resources, and the sea had lost its prosperity. "Restoration of the sea" is the tenet of Setouchi Triennale. To revive local area, architect Tadao Ando and artist Yayoi Kusama programmatically re-establish the islands on Setonaikai, holding this art festival every three year. "The Angel's Path" is a path of sand, which connects four islands, forming a fantasy scenery with islands, sandbars, and the sea. It remains its own tradition, while improving with western technologies and art designs from around the world. It is a traditional island, and a "sea of hope" leading the island to the world as well.

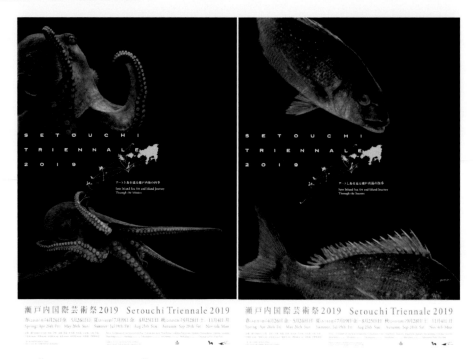

"Undersea creatures" is its theme, hoping to present the unknown dim underworld beneath the calm and silent Setonaikai, gathering those undersea creatures we are not familiar with. Kenya Hara breaks most people's impression, hoping to make viewers feel astonished and amazed, popping up "?" and "!" inside their minds, and at the same time convey the charm of the deep ocean.

d. Circular Economy

Forward-Looking Infrastructure -Circular Economy

CEO of construction company

Kuo-Wen Lin

Taiwan is mainly an outselling country, and ground property is the mother of fortune. While international ground property mostly belongs to the governments, Taiwan has the saying of "ground is fortune" due to its small ground and limited resources. With the citizens' expectation of rising ground property for the future, construction businesses and investors have made use of the opportunity to gain high profit. However, Taiwan has inherited the construction laws from Japan for decades, causing the lack of breakthrough in construction structures, only adopting the constructing styles from western countries. Nowadays, how to make RC construction structures step into the international 3D printing

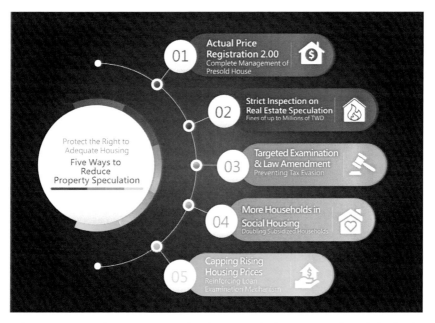

Reference: Yu-Yun Xiao. (2020, December 3). How to Reduce Property Speculation by the Executive Yuan: Promoting 5 Ways to Reduce Real Estate Speculation. https://www. ettoday.net/news/20201203/1868549.htm

area to improve housing situations is an issue, along with general solutions for lack of workers in construction industry and the costs of employees and work quality. Industries should not make profit by deceiving customers. Western countries and China have been implementing 3D housing in the market for years, using more recycled materials (steel slag, scraps, iron cement, etc.). At this time, CSR housing justice is the best chance and challenge for the Construction and Planning Agency to eliminate bureaucracy, continually testing construction structures, implementing and innovating smart city technology, and recreating economic boosts for Taiwan's construction industry.

1. The construction industry has always been the leader of Taiwan's golden age. Aside from leading the consumption of other industries, the construction industry also serves as the indication of city development and environmental development, existing for the relationships and highest value of humanity. For individuals, the word "community" warms people's hearts, but the existence and development of the construction industry depend on the market's needs. Nevertheless, due to the uneven quality of construction companies, they leave a bad impression to people, despite bringing the happiness of homes. The cycling process of construction industries (including planning, production, construction, management, demolition, trading, renting, etc.) and campaigns of industry value chain feature labor intensiveness. Whether the sales and revenues are a success or failure is substantially contingent on the professionalism, and a basic technology group is necessary. The industry properties are easily influenced by government policies and the promotion of important public construction. Because most construction companies are medium or small companies, the construction industry is generally considered lacking in morality and damaging to the environment. Taiwan had begun striving since the end of SARS in 2003, and continued on a twelve years' endeavor. After achieving 14 years' highest records in 2014, prices and quantity both reached high peak, plus opportunists bringing up real estate prices have become the incentive of raising ground prices, which indirectly results in real estates not being able to drive domestic demand, and has become an in-fighting industry. The whole industry made the Taiwan housing black swan event happen, and the government made a policy integrating housing and land tax system in order to lower housing prices.

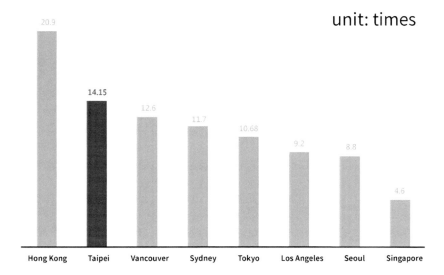

unit: times

Reference: The 2019 International Housing Affordability Survey Report by American public
policy consulting firm Demographia. It does not include surveys in China, Taiwan,
Japan, Korea, etc.; thus, the statistics of Taipei, Tokyo, and Seoul are from the local
governments. The data of Taipei are from the first season in 2019.

2. The influence COVID-19 has does not only show in the economic market, but also
worsens the situation of Taiwan's construction industry by limiting Taiwan from
participating in international organizations. Enterprises are the cells of the society,
and the society is the source of profit for enterprises. The panicking investors flood
to "corporate social responsibility" during the pandemic. CSR is considered the
potential source that has the biggest advantage in competitiveness. It is one of the
most efficient and necessary "operating strategies" in an environment of chaos,
competition, versatility and uncertainty.

(1) Construction industry all over the globe is promoting and implementing
sustainability responsibility, but it is difficult for the domestic construction
industry in Taiwan to advance in the CSR area.

(2) The corporate social responsibility for the construction industry is simply
at the preliminary stage, and the government's and enterprises' lack of
knowledge and activeness for CSR has become a potential problem for
Taiwanese enterprises to compete globally.

(3) Although the construction industry is different from common industries, with complex environment and chains of factors such as laws, environment, cultures, and huge funds, in order not to get eliminated by the economic environment, not only does it have to meet construction quality and safety requirements, but also strengthen technology and creative management ability. If it could be responsible for CSR, it benefits the government, consumers, and the construction industry.

(4) Not to mention the current construction industry is down on prosperity, there are various CSR practices under different cultures, and consumers now value environment-friendly buildings to be integrated. The current construction market has applications, designs on smart buildings and barrier free environment, deduction on burden on the environment and waste of energy, and lowering of the costs of the environment. There are also ISO managing system and CRM customer service system, along with certificate of health quality at home, and ICT Internet of Things to construct the league of delicate value chain and green supply chain, with timber market and SRC steel structured buildings investing in arts one after another......etc. These all require long-term promotion, not achieved quickly, but they are beneficial to adjusting passive attitude.

3. By 2050, infrastructures will grow four times globally, inviting the future trend "the next wave of innovation". I believe 3D printing houses are the only way to solve housing justice. Take western countries for example, even in China, the construction industry implementing 3D printing houses and public constructions quickly solved problems of shortage on workers and quality. If we view 3D printing housing and infrastructure as

the engine of country development, thinking how to innovate technology, it could improve economy, raise employment and domestic demand in the future when the market connects with the globe. Moreover, through integration and flagships, we could connect Taiwan's advantages and chances, create the experiencing and application of life industry, strengthen our competing advantages in Asia and the world, and recreate Taiwan's economic energy in construction industry.

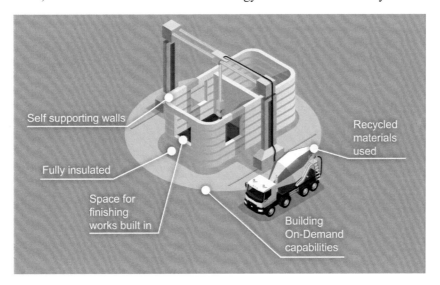

TransAct is an "Anchor Firm": a main firm that prompts the companies related to a supplier's development and creation. TransAct implements integration of various resources, roll plan the policies promoting 3D printing housing in the future. Taiwan's medium and small enterprises and newly created industries all need the support of policies. I have been in the construction industry for nearly 30 years, and I believe that currently there are two prominent points: (1) Laws related to the construction industry being revised to keep up with the world, (2) if the Taiwanese government could pay attention and implement policies of incentives and subsidies, the construction industry could implement this international long-term strategy more. Smart City would solve city problems (the homeless, human resources, population, costs) let citizens be equipped with the ability to buy homes, and resolve the problem of unbalanced supply and demand in housing more comprehensively.

e. Sustainable City, Future City

Sustainable City, Future City

Mayor of Meinong

Bing Guang Jhong

As the leader of the Meinong District, relentlessly creating a better environment for Meinong is not only what the citizens expect of me, but also what I expect for myself.

According to the FAO, the variety of crops has reduced 90% in the past 100 years, which means what we eat now are more monotonous than that of our ancestors'. In recent years, the UN have been promoting the Satoyama Initiative in the hope of shortening the distance between human and nature again, and benefitting biodiversity preservation and sustainability. In order to let farmers in Meinong produce food on the vast plain at the foot of the mountains, there are various traditional crops and rural culture, including the white jade radishes and cherry tomatoes in Meinong. They are the fruit of city and country sharing the hillside agriculture with each other. The energy-saving, insulated buildings are used to reduce the heat in summer and for the change of climate, while the patches of green spaces, roofs, and community plantations all connect to each other, forming an eco-friendly and bio-friendly habitat for creatures in the city. To continue an environment like this, as the leader myself, I strive for resources from the Kaohsiung City Government with all my colleagues in the district. We have built major constructions in Meinong District, for example: the Meinong Lake lakeside environmental facilities, the Meinong Cultural and Creative Center, the Xinwei Bridge, the drainage construction, etc. Preserving the natural habitat of the city while reducing the risks brought by climate change is taking actions for biodiversity. By taking these actions, we also constructed a healthy, sustainable future for the city, meanwhile enhanced both eco-friendliness and the economic development of Meinong, and achieved the goal of building a sustainable city. From the sustainability point of view, the problems and challenges of rural development can be divided into three main aspects: economic, social, and environmental. In lack of local job opportunities, rural areas fall into the vicious circle of rapid outbound migration and a lack of labor force.

Rural cultures are replaced by urban and global cultures, losing the charm and diversity of its local cultures. The agricultural environments are damaged by excessive chemicals. Since biodiversity is the necessary condition for long-term agricultural safety, damaging it would bring danger to global food supplies and food safety. Overall, rural areas contain many aspects like economy, environmental preservation, and cultural activities. To ensure the sustainability of human living condition, solving the social, economic, and environmental problems of rural areas is the most urgent subject right now.

Therefore, aside from increasing infrastructures, we also reinforced community building, social welfare, and meal delivery services to elders living in solitary. From truly reducing poverty and holding celebrations on Father's Days and Mother's Days or fun activities on the Double Ninth Festival, to maintaining infrastructures and offering free village buses for medical attention or business purchasing of remote villagers, we strive to reduce inequality and close the distance of public services between city and country.

Moreover, in order to attract more tourists to Meinong and to make them even fall in love with Meinong, we have held the Colorful Flower Sea Festival in the beginning following a series of special events, giving tourists in Meinong a fundamentally different experience. Besides giving our district the reputation of "the home of smiles", the events have made us into the 2012 top ten Taiwanese village tourist attractions, won us the bronze medal of the 2018 Taiwanese city brands, made Meinong get nominated as one of the 30 most classic villages in 2019, "the year of the villages". These are the achievements of all the members in the district office under the guidance of the city government, as well as the collaborative effort and cooperation of each group and citizen in the Meinong society.

Nevertheless, as the mayor and a native member of Meinong, I will never become complacent about the above results. Aside from the above infrastructures and policies, a village that can attract and keep people must not only be proud and confident of what it does, but also be inclusive and remain open-minded to respect and include groups and people from different places and cultural backgrounds. With this perspective of cultural diversity and the support of the city government, all members at the district office are actively promoting each cultural extension and preservation activity. For example, the district office and several organizations of Meinong (including the Miracle Association and the Guang Shan Temple) hold the cultural activity of "welcome the miracle, pray to words" on the ninth day of the first month of every lunar year to show that the traditional Meinong Hakka rural villages cherish words greatly, and that they respect and desire to explore the sea of knowledge. In addition, the district office actively works with each association, citizen and elder to protect and promote the traditional Hakka humanities and cultures of Meinong, including Hakka Lei-cha, Hakka flat noodles, oil-paper umbrellas, Hakka blue dye, and the making of blue dye clothes. Through the Festival of Yellow Butterflies, the Meinong district office and each association also show how much we care about environmental protection and our effort in preserving a variety of species' habitats.

Besides striving to protect and promote traditional Hakka culture, the district office has been collaborating with the city government, ministries of the central government, and the Syu Sheng Ming Chinese Taipei Baseball Association for years to hold the Syu Sheng Ming Little League Baseball Tournament. By holding the tournament, we hope to continue Meinong people's spirit and common memory for baseball, help Meinong's offspring to inherit Coach Syu Sheng Ming's sportsmanship, and provide the new generation of Meinong's baseball players with a chance to broaden their horizons and to learn from the whole country or even the whole world. Furthermore, it is worth mentioning that the district office also values the living situations and rights of immigrants in Meinong. In 1995, Meinong had already become the first in Taiwan to start "the Mandarin school for denizen brides". After the new immigrants settle in Meinong, the once foreign land has gradually become the home of many immigrants from around the world. In recent years, new immigrants have played a more and more important part in the local economy of Meinong, one of which is the once declining white water snowflake industry. With the effort of a vast number of new immigrants, the industry has not only revived and improved the economic situations of many immigrant families, but also become a local, eco-friendly agricultural representation in Meinong that has gained popularity over the country and has reached an annual output value of over a hundred million. Moreover, the district office has made use of the space in the Meinong Hakka Cultural Museum, and has cooperated with the local immigrant organizations to work on workshops and activities like new immigration, the meaning of moving, and finding a "home". Through these events, on one hand, we help new immigrants alleviate home sick by holding regular meetings, on the other hand, we hope to let new immigrants adapt to life in Meinong and settle in Meinong by interacting with the local community more frequently.

In conclusion, we expect to drive tourism and facilitate business opportunities through these rural agricultural activities featuring rich Meinong Hakka culture. We will make use of the distinctiveness of the local humanities and industry, flexibly apply each aspect of resources with local conditions, keep up with the times, make Meinong a better sustainable city together, and create the vision of future. We will also make Meinong an environmentally friendly place, with respect and inclusion of traditional and diverse cultures. Additionally, it will be a rich village of vigor and vitality, allowing residents to live in peace, and making immigrants from everywhere feel respected, included, and accepted. As the mayor of Meinong, I pray that Meinong will only become better and better, with all of its residents living happily and healthily.

f. New Agricultural Innovation

Urban sustainability combined with new agricultural innovation

Chairman of Dong Shan Farmers' Association

Zhi-Yao Huang

According to the UN's calculation, 1/3 of the total global food produce (equals to 13 billion tons, worth 1 trillion US dollars) is rotten in costumers and retailers' trash cans, or goes bad due to improper transportations or harvest. The food goes into trash can way before goes into your stomach! Meanwhile, there are 50 billion obese population, while almost 10 billion people suffer from malnutrition. Every country has the rights to develop its economic, but the premise is that one should pay respect to the planetary boundaries, (reference: Is the earth really at its limits? Sustainable development can't wait!) ensuring sustainable produce and consume pattern. In this free trade system, the ups and downs of the market price of agricultural products and the impact from climate changes keep lots of farmers and labors from making profits. Many farmers begin their works even before day break, while lots of small-holder farmers earns less than 1,200 NT dollars. They can't afford enough food or the education fee for their kids, and thus long-term live under fear and extreme poverty. The primary goals of modern agriculture in developed countries are to remove the waste on the production side and to solve the commonly overconsumption. In addition, modern agriculture also helps developing countries, improving their skills and technologies, developing production methods in purpose of sustain, providing fair and eco-friendly products and services to encourage and stimulate people's consumption

so as to meet their basic needs and thereby improves the quality of life and productivity.

"Give a man a fish, and you feed him for a day; teach him how to fish, and you feed him for a lifetime." Besides helping small-holder farmer gain self-confident and better abilities, Fair Trade establishes an accountable supply chain and business network with transparent management and business methods so that customers can clearly trace back to the source of the products. Besides, it encourages producers to produce their products in a sustainable way, producing eco-friendly, better and healthier products for customers, establishing the partnership and long-term cooperation between customers and producers, assisting producers and farmers making high quality products and improve produce conditions, cultivating their self-independency. Cities are the major warzone of this global sustainable war. Starting from farmer associations, modern agriculture spreads and covers the whole city, in hope of achieving eco-friendly sustainable cities.

The infinite possibilities of Taiwan agriculture begins in Dongshan Township farmer's association. Circulation designs change and revive local area. Taiwan's first agricultural cultural and creative park begin again from 60-year-old barn in Dongshan!

I'm the director of Dongshan Township farmer's association, and I'm giving the next generation the courage to dream big. We need to provide a sustainable environment. Due to low production, far production areas, lack of market information and else, small-holder farmer often have no rights to bargain. We provide the producers a sustainable way to improve their living quality. Dongshan Township farmers' association, Yilan County, reformed five 60-year-old barns. The once troubled water, Dongshan river, is now a river of hope, mankind and environment's mutual prosperity. Dongshan Township farmers' association hopes and gives their workers the duty of devoting themselves more to Dongshan Township's agriculture. Dongshan Township is an agricultural town. In 2020, the Dongshan Township Farmer's Association reformed the idle barns in front of the Dongshan train station. With the skillful designers' help, the Excellent Food Agri. Park was born within a year. There are four themes: play bar, buy bar, eat bar, and learn bar, all of them have their own features. Based on agriculture and farmers, add on some healthy, innovative, fashionable elements, the park creates a living space which contains traveling experiences, good food, nice presents, and a lot of fun. Excellent Food Agri.

Park is Taiwan's first cultural and creative park featuring agriculture, in there sets special agricultural products promotion area such as small farmers and young farmers area. It advertises that consumers can face-to-face with farmers to shorten the distance between each other, since the park is close to Dongshan river, Dongshan train station, and the old street. It can connect to the Renshan Botanical Garden, Xinliao and Jiuliao Waterfall, and Meihua Lake in the upstream, and 52-jia Important Wetland and Dongshan River Water Park in the downstream. Therefore, the association cooperates with travel industry, hoping to use agriculture as the starting point, planning experience tours, using tourism to improve local economic. The association also showed cooperate intentions with Lion Travel and other travel agencies. They will combine agricultural products and farming experiencing to plan serial tours, hoping to take agriculture as the leader, link up local economics so as to combine with the Dongshan Township Office's "Dongshan River Star" program in the future.

g. The Strategy Models and Writing Essentials of a Corporate Social Responsibility (CSR) Report

The Strategy Models and Writing Essentials of a Corporate Social Responsibility (CSR) Report

Professor of the Department of Commerce and Management, Kao Yuan University

National TTQS Evaluation Committee Officer

Business Start-Up Counselor in the Veterans Affairs Council of the Executive Yuan

Doctor Jie-Lun Li

1. The Mindset of writing a CSR report

(1) The Definition of Corporate Social Responsibility

According to Wikipedia, Corporate Social Responsibility is an ethical or ideological theory aimed to see whether the government, companies, institutions, and individuals have responsibility to contribute to the society. It is divided into Positive CSR and Negative CSR: PCSR indicates that they have a responsibility to participate; NCSR indicates that they have a responsibility to not participate, meaning that a corporation exceeds the expected standards of morality, law, and the public, and takes their influence on each stakeholder into consideration when conducting business activities. The concept of CSR is based on the ideology that business operationsmust conform to sustainability. Aside from considering its own financial and managing situation, a corporation should also take its influence on the society and the environment into consideration.

From the aspect of CSR, stakeholders mainly include shareholders, employees, consumers, the environment, the community, competitors, suppliers, and the government. A corporation has different responsibilities towards different stakeholders. In 1960s, research team from the Stanford University defines that the purpose of a corporation is to serve not only shareholders, but also many essential stakeholders. A corporation cannot survive without their support, and it should try to include factors of the society and the environment. Corporate Social Responsibility Report written to correspond to each stakeholder, or Corporate Sustainability Report, records detail of the corporation's targets, results, promises and plans on sustainability and social responsibility in the form of a report.

(2) The Interpretation of Corporate Social Responsibility

Environmental protection measures such as water and energy conservation, revealing carbon footprints, beach clean-ups, carbon reduction, and waste reduction are just the basics of environmental protection. Nowadays, the practice of environmental sustainability has been integrated into business operation. The purpose of CSR is for corporations to give back to society; they should not only earn profits for shareholders, but also contribute to the sustainability development of the society and the environment. The circular economy that recycles resources into new material is one of the key knowledge in recent years. Examples include DA.AI Technology using recycled bottles to make clothes, and Far Eastern New Century Corporation using sea waste to make sneakers. However, the manufacturing industry is not the only one who can contribute to environmental sustainability. Retailers can also practice environmental friendly methods on plastic reduction, waste reduction, and energy saving. For instance, plastic-free supermarkets already exist in the Netherlands and the U.K., and Carrefour has donated expiring items to reduce food waste and take care of the disadvantaged. To this day, there are still many corporations consider CSR as doing charity, and practice CSR by starting a foundation, sponsoring arts and cultural activities, adopting a park, donating to underprivileged groups, etc. Although these external charitable activities are a part of social responsibility, they are definitely not all of it.

(3) Corporate Social Responsibility and Sustainability

The 17 Sustainable Development Goals of the United Nations include: 1. To end poverty in all its forms everywhere. 2. End hunger, achieve food security and improved nutrition, and promote sustainable agriculture. 3. Ensure healthy lives and promote well-being for all, at all ages. 4. Ensure inclusive and equitable quality education and promote lifelong learning opportunities for all. 5. Achieve gender equality and empower all women and girls. 6. Ensure availability and sustainable management of water and sanitation for all. 7. Ensure access to affordable, reliable, sustainable, and modern energy for all.

8. Promote sustained, inclusive, and sustainable economic growth, full and productive employment, and decent work for all. 9. Build resilient infrastructure, promote inclusive and sustainable industrialization, and foster innovation. 10. Reduce inequality within and among countries. 11. Make cities and human settlements inclusive, safe, resilient, and sustainable. 12. Ensure sustainable consumption and production patterns. 13. Take urgent action to combat climate change and its impacts. 14. Conserve and sustainably use the oceans, seas, and marine resources for sustainable development. 15. Protect, restore, and promote sustainable use of terrestrial ecosystems, sustainably manage forests, combat desertification, halt and reverse land degradation, and halt biodiversity loss. 16. Promote peaceful and inclusive societies for sustainable development, provide access to justice for all, and build effective, accountable, and inclusive institutions at all levels. 17. Strengthen the means of implementation and revitalize the global partnership for sustainable development. The tender of the Ministry of Labor had included CSR in the indicators the earliest, including pay level, pay raise percentage, and work-life balance measures, followed with occupational safety and expansion to other ministries. The social organization innovation that the Ministry of Economic Affairs is promoting also aims to encourage enterprises and organizations to step forward to this goal.

Nevertheless, it requires top-down implementation to integrate CSR into a company's DNA. Top executives must value and participate in CSR in order to influence employees. Otherwise, junior staff are usually busy working, thinking that CSR is only the issue of the human resources department or the public relations department. Thus, only if individuals at the highest level of management truly show their support and promises to CSR can the employees take CSR seriously. Some enterprises have already listed CSR as one of the KPI in performance evaluation. For example, annual reviews for international branch managers of Clarins include 10% of their performance in CSR, and Unilever asks all of its brands to promote at least one sustainability project every year.

2. The Strategy Models of Writing a CSR Report

(1) Three Levels of Planning the CSR Strategy

Nowadays, CSR strategies can be divided into three levels: the first level is corporations participating in charity events and donations, which has no direct relation with their core business; the second level is reducing the company's negative impact on the society and the environment by reinforcing management and improving production processes, at which the company has included external factors in its business decisions; the third level is balancing the collective benefits of the economy, the society, and the environment, and paying equal attention to profit earning and social responsibility. The results are not as expected for plenty of companies, mostly because they did not think clearly about the "social responsibilities" of corporations before planning the specific strategies. They did not find out the proper subject and implementation method of CSR, nor did they include the concept of CSR in the vision of business management.

(2) The Strategy Execution Models of CSR

From the perspective of a corporation, strategy execution models of CSR can be roughly divided into five types depending on its relation with the business development strategies:

a. Resource Assistance Strategy Model:

The company's execution methods include financial assistance, non-designated charitable donations, sponsorship, and general assistance, etc. The resource assistance is mainly one-time or short-term funding of general manpower. This model is not necessarily related to the business's specialty, and CSR is separated from the mission statement and operation of the company. Some of the assistance is even based on the personal will of the business owner. Thus, if the company is taken over by another owner, these assistances could be suspended or changed. Since it has nothing to do with the business's specialty, management could easily develop a biased mentality of "forced donation", which makes it more challenging to build partnership. This assistance model makes up the largest proportion in Taiwan, and sponsorship is the most common within. It includes placemaking project; for instance, TSMC provided financial aid for the renovation of the former Embassy of the United States.

b. Professional Assistance Strategy Model:

The execution methods of this model include the company's employees providing voluntary assistance of professional technology, successful experience sharing, capacity building assistance, etc. Since this model can combine with the business's area of expertise, it can provide further interaction. This model is based upon the business's own specialty, which makes employees more willing to devote themselves, and it could also help the business. Tung Ho Steel Artists in Residence Program is a famous example: each year, Tung Ho Steel invites a Taiwanese artist and a foreign artist to work at its factory in Miaoli and provides them with resources such as scrap steel, professional machines, and technicians during the months they live there. After the creation is finished, Tung Ho Steel holds a reception to exhibit their work. The program's experience has made employees of Tung Ho Steel interested in artistic creation.

c. Resource Matching Strategy Model:

The resource matching model requires the company to be the platform, plan the ideas, appeal and gather various CSR resources, meet the needs of different issues, and match people with different specialty or resources. One of the examples is Sinyi Realty's "Community as One, Community Construction Action Program". Applications from the first stage (2004-2008) mainly started bycontinuing and promoting local industry cultures. The "Youth Returning Home Program" of the second stage (2009-2014) became a rising energy for culture inheritance and sustainability, and afterwards it was found that many communities needed empowerment assistance on application methods, or else they could not complete the application on the matching platform. In 2015, the program was renamed "Community Construction Action Program" and was adjusted to "interactive participation learning". In 2020, the program echoed the Ministry of Education's 2019 curriculum guidelines "interaction and common benefit" to explore community issues and transform the community.

d. Business Cooperation Strategy Model:

The execution methods of this model include social shopping, which drives consumers to purchase products and services with social missions and emphasizes on additional influence. Another method is collaborating with a business partner and planning projectscollectively to provide customized, discounted, or free products.

This model develops CSR strategies from the perspective of business models; thus, the collaborative programs can be joint development, project collaboration, or customer business relations, and they are morepractice-based in contrast to regular theory-based business relations.

e. Strategic Alignment Strategy Model:

The CSR execution methods of this model include coworking and integration of business strategies. The company and the business partner are highly integrated, oreven integrated into a group. They intend to achieve social purpose through CSR execution, so do the departments under the group. The model creates energy for the company's sustainable development, and possesses the concept of both commercial success and social responsibility when choosing customers and managing relationships.

(3) The Strategic Performance Measurement of CSR

The influence of social and environmental issues on enterprises has been increasing, especially when institutional investors are choosing investment targets, they not only consider traditional financial performance, but also take environmental performance and social performance into consideration. Therefore, enterprises revealing non-financial sustainable development report is the tendency of international development. Appeals to standardize the report have been made by various parties, among which the most influential organization is Global Reporting Initiative (known as GRI), founded by the UNEP and the CERES in 1997. The GRI Standards is the only reporting framework in the world co-established by multiple stakeholders, including corporate, labor, human rights, accounting, environmental, and investment organizations. It provides a widely accepted system for organizations to report on their economic, environmental, and social performance. The framework includes sustainability reporting guidelines, various indicator protocols, technology protocols, and occupational supplementary guidelines.

With the era of corporate globalization approaching, more and more people are aware of multinational corporations' performance on sustainability and social responsibility issues. This can be a catalyst for enterprises to make their own way to sustainability and practice CSR as the world hopes them to. Enterprises disclosing their non-financial performance information clearly and elaborately will give them a chance to promise the public specific goals related to their business management and corporate development vision.

3. Practicing the Composition and Execution Steps of the CSR Report

Corporate Social Responsibility is stakeholders-centered, aimed to present the responsibilities of a corporation by making it respond and take responsibility for issues that stakeholders care about, while corporate sustainability is company-centered, aimed to enhance the competitiveness of the corporation by considering the environmental and social aspects of business strategies. Both corporate social responsibility report and corporate sustainability report are communication tools for corporations and stakeholders to consider the corporation's operation in economic, environmental, social, and cultural aspects, and to develop strategies to make the corporation long-standing by using transparent, valid employee development methods. A proper corporate social responsibility report or corporate sustainability report should disclose information adequately and transparently.

With the concept above clarified, when a corporation is ready to compose and publish a report, there usually are four steps:

(1) The first step, implement the systematic framework, policies, and action plans of CSR to confirm declaration method:

What is the purpose of the report? Where are the readers? Why must corporations declare its executive performance of CSR? What report standards should we follow? What kind of CSR information should we declare? How to define and collect CSR-related information? What principles should we follow to declare the information and data of CSR? What are the qualitative features (relevance, authenticity, transparency, etc.)? Different readers would result in different interpretations: consumers would see safe products in a CSR report, college students would see a company's novelty in a CSR report, employees would see company welfare in a CSR report, shareholders would see investment opportunities in a CSR report, researchers would see trend analysis in a CSR report, and NGOs would see partnership in a CSR report.

(2) The second step, clarify major stakeholders, issues they pay attention to, and the content plan of the CSR report:

What messages does the report aim to convey? How to define cases related to CSR? Where are the target readers? What types of stakeholders should be taken into consideration? How to take them into consideration? What CSR information do we aim to declare? What information can we get from management process? How should internal organizations improve to convey important CSR issues, events and indicators authentically? Which systems can we get important information from? How to verify the accuracy of information?

In the past, companies mostly practice short-term, slowly increased CSR programs. Now more and more companies have started to adopt a more comprehensive strategic business model with social mission. Other changes include: integrating CSR with business models, transforming into a business model with social purpose through CSR, adopting data information as the basis of social purpose strategic planning, and actively measuring, tracking, and disclosing the influence the enterprise has on the society. Corporations used to view CSR as a social responsibility that they have to take; now they view CSR as the way to operate an enterprise. Their thought on CSR has also changed from just a bonus to the key of running a business successfully.

(3) The third step, check on the execution performance of business management, sustainable environment development, and social welfare:

Most time spent in this step is for drafting the report and designing the final report. The most important problem at this step is the "data collection period". Since the data collection period of the report is a whole year, most data are collected in the early stages of production. However, integration and analysis usually happen in the final stages; enterprises might misuse the data in the later stages as the basis of their performance in the report, which diminishes the authenticity of it. Therefore, how enterprises observe the effect of data analysis at this stage is fairly important. Moreover, some enterprises that have more experience in report-writing have already started to include third-party assurances to increase the report's authenticity. The point is that the more time an enterprise is willing to spend on data analysis, the higher the quality of the report will be.

(4) The fourth step, review and improve the report after it is published and read:

Spread the information of the report to the previously set readers and stakeholders, including employees of the company, external important stakeholders, NGOs, and financial institutions. The last crucial step is to review and understand the goal of improvement and management direction. After the CSR report is published, the enterprise should have a feedback mechanism actively collecting feedbacks from the public, and let different stakeholders respond to the company's CSR performance. The improvement report should include discussions of long-term and short-term prospects, and the experience should be included in the report production strategy as a reference for the next version of report.

5. Improving the Report Continuously

(1) 5 Key elements to Follow

Cramer, the CEO of the internationally famous non-profit organization BSR, which has been committed to promoting corporate sustainability for almost 30 years, pointed out that in view of stakeholders over the world holding more expectation towards corporations in the future, corporations can follow 5 key elements to publish sustainability reports with more communication effectiveness.

a. The first key element: Compilation of sustainability reports is at an important turning point, and this change may decide the direction of disclosure and composition of the report in the next decades. This change appearing now has a crucial meaning, and it directly highlights the core of publishing the report: Why must enterprises publish the report?

b. The second key element: Enterprises do not publish a report without a purpose. The real purpose of publishing a sustainability report is to encourage action and speed up business transformation. More importantly, an enterprise's sustainability report can also be an excellent reference indicator for investors, civic organizations, and the government; meanwhile, it provides better criteria for a sustainable and equal future.

c. **The third key element:** The report helps business transformation and improves business performance. It helps the company develop tougher business strategies by following the sustainability report framework, attracts more investment, and reinforces business actions.

d. **The fourth key element:** The report brings better sustainable development to the society. It motivates enterprises to form sustainability-specific committees anddevelop management policies for major issues. It also helps enterprises achieve Sustainable Development Goals, the Paris Agreement, and other international human rights indicators.

e. **The fifth key element:** It speeds up policies of equality and sustainable economy. Through the report, enterprises should be able to provide necessary information to stakeholders such as investors, consumers, civic organizations, and the government to speed up the pace of sustainable development on both sides. Only if the company truly follows and practices the framework and guidelines of these sustainability reports can they achieve the goals above. BSR has almost 30 years of experience in the sustainability industry, observing the actions of companies and initiatives, and thinking whether there is undone business, hoping to create a new perspective to guide corporate sustainability reports in the future on how to communicate effectively.

(2) The Innovation of CSR Over Time and the Establishment of an Environmental Cost Accounting System

As times change, business management faces different challenges and innovations at different stages. The 1970s was the decade of cost reduction, the 1980s was the decade of quality, and the 1990s was the decade of service. Under current circumstances, if companies hope to hold the key to success, they should not only understand and satisfy the fickle preferences of customers, but also possess the ability of integration and innovation in order to adapt to the current diversely competitive business environment.

The international environmental management standards ISO14000 was published in 1996, emphasizing that the society has to internalize the environmental cost previously born by the society through assessing product life cycles and environmental regulations. Since then, environmental cost accounting systems were created. To put it differently, in order to achieve eco-efficiency and corporate sustainability management, corporations have to actively improve their product design, manufacturing process, pollution control, after service, and waste recycling throughout the product life cycle. The purpose of corporate sustainability development is to consider the society's environmental cost while earning profits. In other words, its core value is how to transform the power of improving the society and the environment into the rise of business. To achieve this purpose, developing a complete corporate strategy is the first priority. Management and professionals such as accountants can change the way a business works, but whether or not innovative strategies can be implemented depends on the vision and leadership of management. The responsibility of management is to establish and set business goals.

In brief, enterprises must actively improve their product design, raw material selection, manufacturing process, pollution control, after service, and waste recycling throughout the product life cycle along with shareholders, employees, and suppliers. In the process, financial factor is as important as social and environmental factors. The three factors influence the cost of the product or service throughout its product life cycle, operation cost, processing cost, and acquisition cost. Therefore, environmental cost accounting is seen as an important technology in achieving corporate sustainability development. The goals are: a. To make promise to sustainability development. b. To help ensure sustainability development is embedded in an organization activity. c. To demonstrate how an organization executes, to show the organization's promise and develop quantified and time-bound sustainability development goals.

h. B Corp.

CSR practical business model strategy – taking the B Corp model as a reference

Managing Director of Taiwan Impact Investing Association

Da-Wei Zhang

CSR is a famous doctrine of today's international companies, and it is also an important condition for multinational companies to choose industrial supply chain partners, thus extending the ESG (sustainable management) evaluation model. In fact, including early CSR, to later social enterprises, corporate governance, the Equator Principles (EPs), carbon footprint and green finance, are also some new principles arising from the continuous evolution of CSR. CSR in Taiwan listed companies and some non-public offering large companies has done a good job, listed companies in some industries are required to publish CSR reports every year, and although there is no absolutely fixed format due to different industrial characteristics, the reports of the various companies are similar. However, CSR is mainly an evaluation specification designed for large enterprises, and the production cost is high. Therefore, even if the listed companies are integrated, it is still not suitable for the evaluation of the majority of Taiwan's SMEs, and even micro-enterprises are completely incapable of making CSR reports. But as B Corp model are concerned, it is another CSR evaluation mechanism, and it can be fully applied to both large enterprises and SMEs, even micro-enterprises, as long as they have the will, they can also do the projects that required by the evaluation requirements of B Corp model. I would not say that B Corp model are modified CSR or simple CSR, because it is completely an independent evaluation standard, but it is fully in line with the major principles of CSR. Basically the application is not difficult, is not complicated (probably every two business owners on different complex definition), which requires companies to apply only to go personally, and re-examine every three years, to make sure he did not leave the track of B Corp model, so that enterprises can comply with the spirit of sustainable operation of it. Therefore, B Corp model emphasis on corporate daily operations in a natural habit to do something, and create a new profit-making use of the process of doing these things in business or operating mode style to profit.

In the following, I will use my personal experience in operating the China Credit Information Service Ltd., the process of becoming a co-founder of B Lab Taiwan, and the experience of leading China Credit Information Service Ltd. to become a B-Corp, to share with companies and readers who care about CSR development. First of all, I have to say when I step into B Corp model, I am not a leapfrog evolution, but I have been thinking about what subjective and objective conditions does a good company needs for a long time. Because I have long served as general manager in China Credit Information Service Ltd., our company are doing the enterprise credit investigation and daily corporate rate every day, in addition to the overall economic level and the industry cycle, the company's founding time, company background, operating status, financial status, the letter of the person in charge of the company, positive and negative news, whether there are patent applications, upstream customers, downstream suppliers, the situation of related companies and so on are all included in the scoring items of the credit report, but I always feel that some factors are still missing! Happened twenty years ago, I started to get in touch with CSR. As a corporate operator, I personally learned that in addition for a good company to make profit, it is also important that what kind of model it choose to make profit. At the same time, business operations are taken from society and also used in society, so they should bear "Corporate Social Responsibility." Even if a company can make money, it is also very important to use what kind of profit model to obtain. (For example, if a company illegally dumps waste and drains at night to reduce costs and earn illegal gains, and then uses the money for charity to atone for the crime, it will still be an illegal business after all.) When we worked with Himalaya Foundation to study the evaluation structure of "Corporate Social Responsibility", Australian CSR research institutions visited Taiwan and told us that Australia has launched some evaluation mechanisms to package CSR ETF portfolios and studied. I found that the constituent stocks included in the portfolio are all highly profitable companies, from which I realized the positive importance of the link between CSR and corporate management. I wanted to incorporate CSR evaluation into one of our credit report scoring projects, but at that time Taiwan's corporate governance was not very mature, the CSR environment had just started, and the CSR evaluation project was complicated. If you want to put it in a credit report in the report, the time and cost of the report were very expensive, so the plan to introduce the credit report was shelved. However, we added the item of fines for corporate environmental factors to the credit report and became one of the contents of the report. At the same time, I did not give up the idea of pursuing CSR. I'm promoting the "green Earth Day" in our company by myself, to make colleagues do environmental charity with company during holidays. Organize second-hand merchandise sales in the company and donate the proceeds from the sales to charities. Encourage colleagues to donate their daily income together with the company to provide donations to disaster-stricken households when a major disaster occurs in the country. But I think it's not enough just to do charity. So in 2014, I started to study "Social Enterprise" again to understand the purpose of it as to solve a certain social problem. In order to solve this social problem, operating income can be obtained by designing a service or product through a fee, and this operating income can be used to deal with the social problem to be solved. We have organized a number of social enterprise seminars and forums, inviting operators to share their development and successful experience with our customers and financial industry, with the purpose of promoting the concept of social enterprise.

It seems that social enterprise is a bit similar to the current trend of impact investment, but social enterprise is not the main purpose of making profit, and it is a completely different concept from impact investment. In the process of getting in touch with social enterprises, it was only in 2015 that we really came into contact with "B Corp model". So what exactly is a B Corp model? In short, it means that after meeting certain standards in the five major aspects of environment, employees, customers, communities, and corporate governance, and passing the certification of the American B Lab, you can become a member of a B Corp model. In other words, the B Corp model is not advertised as a social enterprise, but on advertised that an enterprise must not only be profitable, but also in the process of operating, uphold the moral conscience of the enterprise, and create a peaceful and well-meaning enterprise within its own capabilities. Public welfare enterprise and social environment.

Since B Lab issued the first B Corp model certification in 2007, there have been more than 3,600 B-Corps in the world, but their proportion in the total number of global companies is still very small. There are currently 29 B-Corps in Taiwan, and the number locate forefront in Asia. However, the proportion of more than 1.62 million enterprises and trade names in Taiwan is still less than 0.002%, which shows that such a concept of a "good enterprise" requires a strong force to promote it. Therefore, how to "bring the world's B-Corps into Taiwan" became the driving force for me and my three partners, Lian, Ting-Kai, Hu, De-Ci, and Chen, Yi-Ciang to establish the B Lab Taiwan in 2016. Through our own experience of becoming a B-Corps, we influence and drive other companies to follow up. Even if they didn't actively apply for the B Corp model certification, they can still become a good company that meets the B Corp model. As mentioned earlier, the "B Corp model" emphasizes that the enterprise must do what it takes. So it is necessary to have an in-depth study in environment, employees, customers, community and corporate governance. Fill out the questionnaire through B Lab's official website to check what the company itself has done to meet the B Corp model indicators. Not only do they feel that they have done it, but they also need to be able to show that they have done it with actual information. For projects that have not been achieved in the five aspects, it is necessary to think about how to improve to meet the certification requirements. Not only must the company itself have the power to achieve it, but it must also inspire its employees to be willing to cooperate and develop daily habits.

For example, when we applied for B-Corps, we were looking at how we could save energy. I have the habit of turning off the lights during noon break. In summer, the air-conditioning is maintained at 28 degrees. The office area uses LED lights for lighting, and the old ones are replaced by LED lights. And use each utility bill to review our energy saving results. For us paperless work is not easy, but we were already going on this direction to improve the computerized process designed deepen, and data review of various investigations has been changed to electronic review. Although there is no way to achieve a 100% paperless operation, through the management of business machines, we can understand which departments or colleagues have significantly reduced the amount of paper used. This is all practical evidence and can be submitted in writing when applying for applied for B-Corps. For lunch in the company, no wash-free chopsticks and bowls are completely eliminated. Colleagues are required to bring their own environmentally friendly tableware, and even design environmentally friendly chopsticks and straws to send to customers, hoping to take practical actions to influence customers to do environmental protection with us. These are all bonus points when applying. For the sake of employees, we strictly prohibit employees from working overtime. We hope that employees can replace overtime with efficiency, leave work on time, and enjoy weekend breaks with their families. And we have taken the initiative to increase salaries for three consecutive years, and I have taken the initiative to design the employee dividend system. And I am willing to use the company's surplus to provide employees with a higher year-end bonus than in the past. Before I stepped down as the general manager in 2017, I offered employees the highest year-end bonus compared to the past for 4 consecutive years. This is not because I have to apply for a B-Corps. but a B-Corps. has made me realize that employees are valuable assets. In my 30-year career as a general manager, we have almost never lost money, and we have all issued year-end bonus, but in the spirit of the B-Corps. I am more willing to share the results of the company's operations with employees. I hope that my colleagues who work hard together will agree with my philosophy and the company as a big family. I realize the joy of being a happy enterprise. These are all the experiences I gained after personally experiencing when running a business.

Of course, the first condition of a B-Corps is that the enterprise must be profitable. And use different business models to run the enterprise. It is not to squeeze out profits by reducing costs (I did not say that reducing costs is wrong, but emphasized that enterprise should not rely on reducing costs to achieve profitability, because reducing costs is something that should have been done.). Therefore, we strive to automate the operation and save the operation time of the credit report in an automated mode, including automated modules, automated data comparison, debugging, automated forecasts, and simple automated reports. In this regard, we have transformed very quickly and well. In the use of the database, we cut the data more refined but also easier to query. Therefore, we have changed the basis of customers and put our vision on international customers. Therefore, we do not need to deliberately reduce costs to generate profits, but increased profitability because we change the operating mode. I found that from the moment we were willing to start, we changed the company together with our employees and couldn't stop. What makes me more gratified is that we no longer need to prove that we are better through evaluation, because we ourselves will have the idea of wanting better change. We also build a B-Corps ecosystem through the B Lab Taiwan. We try to source our procurement through partner companies that are also B-Corps. At the same time, we also require suppliers to do it with us to expand the influence of the B-Corps. At that time, ten companies, including JIA WEI & Co., CPAs, Ming Yung Enterprise Co., Ltd, and Chia Pang Plastics, were the pioneers of B-Corps in Taiwan. After obtaining the B Corp model certification, what changes have been made to internal employees and corporate governance? It has a positive impact on external customers, suppliers, and even the society and the environment. What we have in common is that we are all SMEs, and we are also willing to challenge the certification of B-Corps with the perseverance and spirit of "taking the bull by the horns" and "to dream the impossible dream", and reinterpret the different business vision and future of enterprises.

Enterprises reflect the awakening of the "green business", "social enterprise", "sustainable business", all the way to the evolution of "B-Corps" birth, but in the attitude and actions "B-Corps" is compared to the previous kind of business type to be more a more active one type of business, with new strategies to achieve the purpose of business survival, but is the more profitable business! The reason is that they have changed their business ideas, have the courage to assume corporate responsibilities, and dare to do things that others dared not try, so the entire business model has also changed. I believe that no matter how large or small a company is, as long as it is committed to practice, every company can become B-Corps. Even it cannot pass the B Corp model certification, as long as it is willing to take on a little more social responsibility, extend a pair of benevolent hands, and save more. a little social good side, they can make better business itself, but also allows the whole social atmosphere more than a one point better. The B-Corps shapes a model of current and future business operations. In terms of business philosophy, "sincerity" is the value, and "Put yourself in someone's shoes" as the starting point. My personal feeling along the way is that B-Corps are the easiest CSR microcosm for small and medium-sized enterprises. Therefore, when Taiwan wants to build a CSR social concept, it will be a top priority to build a CSR evaluation model for Taiwan's SMEs by relying on the B Corp model. At the same time, I also call on SMEs to move on their own. No matter which CSR evaluation model they participate in, they must be physically awakened. As long as they start to do it, the enterprise will change and become better.

Reference

- CSR for Culture. (2020, June 19). CSR for Culture: Cooperation Between Corporations and Cultures. https://csr.taicca.tw/csr-for-culture-%E4%BC%81%E6%A5%AD%E8%88%87%E8%97%9D%E6%96%87%E7%9A%84%E5%8D%94%E5%8A%9B%E6%A8%A1%E5%BC%8F

- Jun-Tang Wang. (2020, December 10). Director Gave Up High Salary to Open a Book Store. https://www.businesstoday.com.tw/article/category/183035/post/202012100011/%E6%9C%80%E5%BC%B7%E9%A4%A8%E9%95%B7%E6%A3%84%E7%99%BE%E8%90%AC%E5%B9%B4%E8%96%AA%EF%BC%81%20%E7%A0%B8%E5%8D%83%E8%90%90%AC%E9%80%80%E4%BC%91%E3%80%80%8D%EF%BC%81

- Central Taiwan Science Park Corporate Sustainability. How to Write a CSR Report? http://www.ctspcsr.com.tw/article_list/view_article_detail/?id=41

- Executive Yuan. (2019, January 30). Circular Economy Promotion Program. https://www.ey.gov.tw/Page/5A8A0CB5B41DA11E/18ef26a4-5d05-4fb3-963e-6b228e713576

- Executive Yuan. (2020, March 11). Foresight Infrastructure Project: Establishing the Foundation of Country Development 30 Years in the Future. https://www.ey.gov.tw/Page/5A8A0CB5B41DA11E/9cf2eef1-e2d2-4f37-ba6e-9498deb422b4

- Zi-Xian Wu. (2020, October 21). 210 Thousand Pieces of Garbage Per Square Kilometer! 8 Dirty Spots on the West Coast of Taiwan, with a Density of Marine Debris 1.5 Times as Much as that of the Global Average. https://www.businesstoday.com.tw/article/category/80392/post/202010210015/%E6%AF%8F%E5%B9%B3%E6%96%B9%E5%85%AC%E9%87%8C%21%E8%90%90%AC%E4%BB%B6%E5%9E%83%E5%9C%BE%EF%BC%81%E5%8F%B0%E7%81%A3%E8%A5%BF%E6%B5%B7%E5%B2%B88%E5%A4%A7%E9%AB%92%E9%BB%9E%EF%BC%8C%E6%B5%B7%E5%BA%95%95%E5%9E%83%E5%9C%BE%E5%AF%86%E5%BA%A6%E6%98%AF%E5%85%85%A8%E7%90%9031.5%E5%80%8D

- Yan-Ting Wu. (2015, December 20). The Furthest Distance in the World is When the Food We Cannot Finish Goes to the Trash Can Instead of People Who Need It. https://www.seinsights.asia/article/3289/3271/3748

- Pei-Xuan Lin. (2017, August 18). 20 FAQs of CSR. https://www.gvm.com.tw/article/39488

- Pei-Xuan Lin. (2017, September 29). 20 Certified B-Corps in Taiwan, the Biggest Number in Asia. https://www.gvm.com.tw/article/40287

- The Forestry Bureau. (2020, April 29). The Strategic Framework of Promoting the Satoyama Initiative in Taiwan. https://conservation.forest.gov.tw/0002057

- Weplus Technology. The Air Exchange Natural Transform System. http://www.we-plus.com.tw/productsType.php?ProductsTypeID=05
- The Rural Meinung Field Learning. https://meinungfield.wordpress.com/%E9%97%9C%E6%96%BC/
- Zi-Rong Chen & Yi-Jing Wu. (2020, June 8). 2020 World Oceans Day: *The Beach Cleanup Manual*. Cleanse the Ocean; Respect the Sea. https://e-info.org.tw/node/224927
- Yi-Hui Zhang. (2021, January 7). Director Put All of His Retirement Pension into Offering Free Tutoring for 10 Years. https://www.chinatimes.com/newspapers/20210107000596-260107?chdtv
- Kai-Ping Ye. (2018, June 25). Go Beach Cleaning for Once. https://www.greenpeace.org/taiwan/update/1765/%E5%8F%AA%E8%A6%81%E8%A6%AA%E8%87%AA%E6%B7%A8%E7%81%98%E9%81%8E%E4%B8%80%E6%AC%A1/
- Jun-Ping Zheng. (2008). *A Manual Guide to Corporate Social Responsibility*. Taipei:Commonwealth Publishing.
- Bureau of Agricultural, Tainan City Government. (2014, November 5). Farmers' Association Alliances. https://agron.tainan.gov.tw/News_Content.aspx?n=1262&s=150588
- Jia Wei CPAs. The B-Corp Movement: Make Businesses Become the Power to Change the World. https://www.jwcpas.com.tw/publicationsBook3-show2.php?book3p_id=236
- Wikipedia. (2020, September 12). Aquatic Ecosystem. https://zh.wikipedia.org/wiki/%E6%B0%B4%E5%9F%9F%E7%94%9F%E6%80%81%E7%B3%BB%E7%BB%9F
- Wikipedia. (2021, March 13). Corporate Social Responsibility. https://zh.wikipedia.org/wiki/%E4%BC%81%E6%A5%AD%E7%A4%BE%E6%9C%83%E8%B2%AC%E4%BB%BB
- Jing-Hui Liao. (2016, December 19). Agriculture Businesses are Profitable! Meinung Farmers' Association Takes the Initiative of Practicing the Satoyama Spirit. https://www.thenewslens.com/article/56889
- Yi-Chen Zhang. (2021, January 16). Titled the First Agricultural Cultural and Creative Park in Taiwan, the 60-year-old Dongshan Barn is Innovated. https://udn.com/news/story/7328/5179012
- Taiwan Stock Exchange Corporate Governance Center. (2021, February 3). Introduction to Corporate Social Responsibility. https://cgc.twse.com.tw/front/responsibility
- Hao-Ping Lu. (2018, December 18). The Sea is Suffocating! 80% of Marine Plastics in the World are from Asia. https://www.gvm.com.tw/article/55332
- UDN Vision Project. (2020, June 9). "If the Ocean can Count on You" Special Reports. https://today.line.me/tw/v2/article/QnKjr0
- He-Zheng Yan. (2019, January 3). What is Corporate Social Responsibility? Understand CSR, ESG, and SDGs. https://csr.cw.com.tw/article/40743
- Aron Cramer & Dunstan Allison-Hope (author). Yi-Cheng Shi from CSRone (translator). (2020, December 1). BSR: The 5 Key Elements for Future Sustainability Reporting. https://csrone.com/topics/6569
- Bowen, H.R. (1953). *Social Responsibilities of the Businessman*. New York: Harper & Brothers.
- CSRone & NCCU College of Commerce Sinyi School &PricewaterhouseCoopers Taiwan. (2017, March). Taiwan's CSR Report Survey in 2017. Taipei: Co-published by CSRone & NCCU College of Commerce Sinyi School &PricewaterhouseCoopers Taiwan.
- European Union (EU). (2021, February 3). A Renewed EU Strategy 2011-14 for Corporate Social Responsibility. https://eur-lex.europa.eu/legal-content/EN/TXT/?uri=CELEX:52011DC0681
- Global Reporting Initiative (GRI). (2016). *Sustainability Reporting Guidelines: Version 4.0*. Amsterdam: GRI.
- Jessica Seddon et al. (author). CSRone Hope Wang (translator). (2019, July 9). 5 Under-recognized Impacts of Air Pollution. https://csrone.com/topics/5631
- KPMG. (2021, January 12). 2021 KPMG: CSR Has Changed; Social Innovation is the New Favorite. https://home.kpmg/tw/zh/home/media/press-releases/2021/01/tw-csr-social-innovation-seminar-2021.html
- Levitt, T. (1958). The Dangers of Social Responsibility. *Harvard Business Review*, 36, 41-50.
- PChome Online. (2020, June 7). The Ocean is the Womb of Life. https://mypaperapp.m.pchome.com.tw/carled31599/post/1380326363
- Searcy, C. &Buslovich, R. (2014). Corporate Perspectives on the Development and Use of Sustainability Reports. *Journal of Business Ethics*, 121: 149-169.

TransAct is a good global partner that assists companies in converging CSR practices and integrating GRI standards into corporate daily activities and business strategies in the short, medium and long term.

BZA0001

薪傳智庫CSR實務商模戰略

國家圖書館出版品預行編目資料

薪傳智庫CSR實務商模戰略/薪傳智庫數位科技股份有限公司
著. -- 初版. -- 臺北市：藍海文化事業股份有限公司,
2021.05
　面；　公分
ISBN 978-986-06041-1-5(平裝)

1.企業社會學

490.15　　　　　　　　　　　　　　　110007088

版次：2021年5月初版一刷

作　　　　者	薪傳智庫數位科技股份有限公司	
發　行　者	楊宏文	
總　編　輯	蔡國彬	
責　任　編　輯	林瑜璇	
封　面　設　計	余旻禎	
版　面　構　成	徐慶鐘	
出　版　者	藍海文化事業股份有限公司	
地　　　　址	100003臺北市中正區重慶南路一段57號10樓之12	
電　　　　話	（02）2922-2396	
傳　　　　真	（02）2922-0464	
購　書　專　線	（07）2965-267轉236	
法　律　顧　問	林廷隆　律師	
	（02）2965-8212	
版　　　　權	藍海文化事業股份有限公司　版權所有・翻印必究	
	Copyright © 2021 by Blue Ocean Educational Service INC.	
聲　　　　明	本書所有內容未經本公司書面同意，不得以任何方式翻譯、抄襲或節錄	

教學啟航 · 知識藍海

藍海文化

Blueocean

www.blueocean.com.tw